RADIATION PROTECTION

ICRP PUBLICATION 17

Protection of the Patient
in Radionuclide Investigations

A Report prepared for the International
Commission on Radiological Protection

Adopted by the Commission in September, 1969

D1409462

PUBLISHED FOR

The International Commission on Radiological Protection

BY

PERGAMON PRESS

OXFORD · NEW YORK · TORONTO
SYDNEY · BRAUNSCHWEIG

Texas Tech University
School of Medicine
Library

7847

Pergamon Press Ltd., Headington Hill Hall, Oxford
Pergamon Press Inc., Maxwell House, Fairview Park, Elmsford,
New York 10523
Pergamon of Canada Ltd., 207 Queen's Quay West, Toronto 1
Pergamon Press (Aust.) Pty. Ltd., 19a Boundary Street,
Rushcutters Bay, N.S.W. 2011, Australia
Vieweg & Sohn GmbH, Burgplatz 1, Braunschweig

First edition 1971
Library of Congress Catalog Card No. 70 163853

This is one of a series of reports prepared as background material for the International Commission on Radiological Protection. These reports, published in blue covers, form part of the Commission's continuing review of information intended to provide scientific bases for its Recommendations, which are published in brown covers. The Commission hopes that the publication of the reports in blue covers, while not necessarily implying recommendations for present action, will stimulate discussion on matters having direct relevance to its work and to the development of the fundamental principles of radiological protection.

Printed in Great Britain by A. Wheaton & Co., Exeter

08 016773 X

M ef
WN
650
In 8
no. 17

Contents

Preface

IN 1968 the Commission asked Dr. R. E. Ellis to prepare a report on protection of the patient in radionuclide investigations. The report describes the basic principles for minimizing the dose to patients receiving radiopharmaceuticals, and also presents a compilation of estimates of the absorbed doses resulting from the administration of the more commonly-used pharmaceuticals.

The report is complementary to the report, prepared by a task group of ICRP Committee 3, on *Protection of the Patient in X-ray Diagnosis* (ICRP Publication 16).

Protection of the Patient in Radionuclide Investigations

A. Introduction

During the last few years there has been a large increase in the number of patients who have been investigated using the widening range of radiopharmaceuticals available clinically. There is an increasing choice of alternative methods, from the well-established ones, using well-known nuclides, to those methods utilising new, and, in particular, short-lived nuclides. In consequence, some clinicians are unfamiliar with the radiation doses received, particularly from these new nuclides, and of the dose commitment to organs in which the highest concentrations occur. Appreciable doses to organs may arise from low concentrations with long retention of particular nuclides. There are also quite large gaps in metabolic information, making accurate estimates of dose difficult.

In the literature, the doses estimated for some investigations show quite large discrepancies. There is thus a need for a reassessed listing of estimated doses and also for a commentary on the protection of the patient undergoing investigations with radiopharmaceuticals. This review indicates the factors involved in minimizing the radiation dose necessarily received during diagnostic tests with radionuclides.

The listing has been given in terms of the dose in millirad per microcurie (mrad/μCi) administered for nuclides and pharmaceuticals in clinical use. Where the investigation has become reasonably well accepted, the typical activity in microcuries (μCi) ordinarily administered is also given. These data then allow an estimate of the dose to be calculated for a typical investigation. However, there will still be a considerable variation in the absorbed dose per microcurie, depending on individual metabolism within the normal range, apart from variations in individuals with abnormal metabolism.

These listings must be interpreted in the light of the clinical need, which must determine the precision required, whether in localisation or in quantitative uptake, and the number of repeated tests that are required. Thus, no absolute dose limit is appropriate, but in each case the need for information must be judged in the light of the risk entailed by the investigation itself. Nevertheless, it is still important to review the size of the dose commitment from a particular radiopharmaceutical and to compare it with that from alternative tests using other radiopharmaceuticals, and also to compare the resultant radiation hazards with the hazards from tests not involving irradiation of the patient which give equally adequate information. A selected bibliography pertinent to the dosimetry of the nuclides has also been included.

B. Choice of Nuclide and Compound

There is a variety of ways in which the same clinical information may be obtained and it is important that the method used should minimize the radiation dose received in obtaining the necessary information. The available methods may include the use of external x-ray, as well as of particular radiopharmaceuticals. (A listing of the doses received from x-ray investigations has been compiled by the ICRP Task Group on Protection of the Patient in X-ray Diagnosis [1].)

The particular properties of nuclides or pharmaceuticals that may determine which is the best for a particular investigation are:

1. Half-life.
2. Energy of emission.
3. Type of emission.
4. Chemical form.

1

5. Availability of substitute elements or compounds.
6. Requirement of test.

1. Half-life

It is important that an adequate counting-rate or detection-rate is obtainable at times biologically appropriate for the particular measurement required. Considerable activity should not persist beyond this time, nor should undue decay have occurred before the measurements can be started. There will thus be an optimal relationship between the half-life of the nuclide and the time of the test. A possible criterion is that the duration of the test should be equal to the effective mean-time for the activity in the limiting organ or tissue (the mean-time being 1.44 times the effective half-life). For example, where a 4-hour thyroid uptake is the appropriate test, then ^{132}I is much preferable to ^{131}I, but where a 48-hour protein-bound radioiodine is required, then ^{131}I is clearly preferable to ^{132}I. Similarly, the use of ^{197}Hg-labelled neohydrin in brain-scanning at a few hours after administration gives a smaller dose for a given counting-rate than occurs when ^{203}Hg is used as the labelling nuclide.

The half-life of the nuclide, however, obviously limits the length of time for which labelled compounds and pharmaceuticals can be stored. Decomposition may occur from self-irradiation, and hence for a particular specific activity of a compound, the longer the half-life the greater the dose that the compound will receive, and thus the greater possibility of causing significant radio-chemical changes. These are known to occur in compounds labelled with long-lived ^{14}C when they have been stored for a considerable period.

When the clearance rates of organs are fast, a high activity needs to be administered so that an acceptable number of disintegrations is recorded by the detectors. To counterbalance the high activities, short-lived nuclides are highly desirable for these investigations, to reduce the dose received. As the reduction of the effective half-life is the important criterion for the reduction of dose, the reduction of the biological half-life is also important. Therefore radiopharmaceuticals having shorter biological half-lives in the organ of interest will also contribute to the lowering of the dose received as long as the biological half-life is long enough compared with the period of the test.

2. Energy of Emission

The energy of the emission is an important factor in determining (1) the dose delivered, (2) the activity required to carry out a particular investigation, and (3) the design of the collimator used with the detector.

The dose delivered to an organ will increase proportionately with the β-ray energy of the emission. For a photon-emitting nuclide, even though the absorbed fraction will decrease in a particular organ with increasing energy, the actual energy absorbed in a large organ will increase.

When the uptake of a particular organ needs to be estimated, the optimal energy of emission is dependent on the depth and thickness of the organ. The requirement is to have a maximum counting-rate in the detector from the activity in the organ, but not to detect as efficiently activity within the tissues lying more superficially or deeper than the organ. A radionuclide having an optimal energy of emission will be one therefore for which the emitted radiation will be (a) substantially attenuated by depths of tissue greater than the depth of the organ of interest, and (b) adequately defined by a reasonably-sized collimator which will lead to the selective detection of radiation arising within the organ of interest and to the attenuation of radiation arising in the surrounding tissues.

It has been shown that for optimum organ scanning, a nuclide emitting photons only of energy in the range 100–200 keV is required. ^{99m}Tc, ^{113m}In and ^{75}Se are of current interest for scanning purposes, as they have emissions in this range. It is possible to build focused collimators having good shielding properties at these energies.

In autoradiographic studies the resolution of the photographic image will be increased by the use of lower energy β-emitting nuclides. The divergence of the β particles into the emulsion from a particular labelled volume of the section will depend on the range of the β rays, since this range is directly proportional to the energy of emission.

3. Type of Emission

As has been stated in the previous section, photon-emitting nuclides are preferred for organ scanning or uptake measurements because with no β-emission, the doses to the organs are substantially reduced for a particular number of photons counted at the detector.

Accurate localization of uptake in organs can be carried out with positron-emitting nuclides by the use of coincidence-counting techniques to detect the annihilation radiation. ^{74}As, ^{64}Cu and ^{18}F are examples of positron emitters used for this purpose in clinical medicine.

4. Chemical Form

Initially, most of the radionuclides were used in simple forms for clinical investigations. Recently, efforts have been made to develop pharmaceuticals that concentrate in particular organs. These pharmaceuticals can be synthesized with the most suitable radioactive label incorporated. This then has altered the whole problem of the retention of the radionuclide. Initially, the metabolism of the ion itself had been considered, but now an understanding of the more complex metabolism of the pharmaceutical is required before an estimate of the radiation dose can be made. The use of column isotope generators for the production of solutions of short-lived daughter radionuclides has also been introduced during the last few years. Yet the loading of the column with the longer-lived parent nuclide must be such that under no circumstances can the

parent nuclide be eluted from the column to contaminate the short-lived eluate. Similarly, if there were non-radioactive "breakthrough" products in the eluate, as, for example, the presence of zirconium in the eluate from ^{113m}In generators, there would be a chemical toxicity hazard.

A large number of compounds have been labelled with tritium and carbon 14. Even though the metabolism of some of the simple compounds is well known, a very large number of compounds labelled with these particular nuclides have poorly understood metabolism. Sometimes, in fact, the whole purpose of the labelling is to investigate and determine the metabolic pathways. In many of these cases, 70% to 80% of the administered compound and nuclide may be accounted for, but the metabolism of the remainder is unknown.

As more pharmaceutical agents are synthesized, there is the possibility of being able either to enhance or to block the uptake of a particular radionuclide or compound into certain tissues. If the uptake is enhanced in the organ of interest, less activity needs to be administered, and the dose to other organs will be reduced. If, on the other hand, some organ on which measurements of uptake are not required would otherwise receive a high dose, its uptake may be blocked by choice of a suitable drug.

5. Availability of Substitute Elements or Compounds

It is sometimes possible to use particular tracer nuclides as substitutes to follow the metabolism or distribution of other nuclides for which convenient radionuclides are not available. For example, radioactive bromine may be used to estimate chlorine spaces, and krypton to estimate the uptake of oxygen. Similarly, pertechnetate (containing labelled technetium) behaves like iodide in entering the thyroid and some other glands, although it does not become retained by organic binding in the thyroid gland, as iodine does.

The choice of the label for pharmaceuticals

may be based on one of the previously discussed considerations in those cases where the label does not influence the chemical properties. However, it is very important that the label must remain adherent, and, if it is shown that metabolism of the compound occurs, as in iodine-labelled human serum albumen, or if the label becomes detached, as the chromium does from tagged blood cells, then it is essential that due consideration be given to the metabolism and organ of concentration of the released activity.

The localization of some pharmaceuticals in particular organs arises not only from the selective metabolic absorption, but also from the increased permeability of some tissues to particular pharmaceuticals, such as occurs in the concentration of neohydrin in cerebral tumours.

6. Requirement of Test

The main divisions of investigations are:
(a) Uptake or dilution in a particular organ.
(b) Clearance rates of organs.
(c) Scan of organs.

A consideration of the particular type of investigation shows that the choice of nuclide is dictated by the speed with which the uptake in the organ occurs when considering the time after the administration at which an uptake or scan needs to be performed. When a measurement of a turnover rate is being made, the half-life of the nuclide must be sufficiently long compared with the duration of the turnover period. Similarly, in dilution studies the time taken for equilibration through the body spaces must be taken into consideration when choosing the nuclide to ensure that sufficient activity is present at that time.

The use of sufficient activity to obtain accurate measurements of clearance rates has already been commented on in section B.1.

The design and type of detecting equipment used for a test should be that which has maximum sensitivity and therefore allows the minimum activity of a particular nuclide to be administered. For example, uptake measurements with scintillation counters usually allow lower activities to

be administered compared with those when Geiger counter detectors are used. The administration of unnecessarily high activities should not be condoned because of the continued use of inadequate equipment, if this can be avoided; however, in some cases, this may be inevitable. Likewise, in some cases, the available facilities may quite properly determine the choice between alternative tests involving comparable doses for equal information.

C. Organ of Reference

In any particular investigation the "organ of reference" will be that organ the dose to which is of greatest concern, from a radiation hazard viewpoint. Frequently, this will be the organ receiving the highest dose. Information as to which organs are likely to receive the higher doses is therefore required, so that any limitation of dose, and hence of administered activity, may be considered. A knowledge of the total dose in rads received during the decay of all the administered activity, i.e. the dose commitment, is, in any event, required. This dose will depend on the metabolic factors involved and also on the route of administration.

The assumption that the organ of reference will be the organ receiving the highest dose may not always be correct. For example, in the case of oral administration, some portion of the gastro-intestinal tract may receive the highest dose, but one will also be concerned with the organs which accumulate the greatest dose from the activity absorbed into the blood. Again, especially in young people or pregnant women, the dose to the gonads or to the foetal gonads may be the determining factor in the amount of activity administered.

Problems often arise in the calculation of the dose received by the organ of reference because the physiological parameters are not accurately known, and only approximate values of organ uptake are known. In some cases, these have to be deduced from information on similar chemical substances. For example, when considering nuclides in the blood, the dose to the blood itself or to the bone marrow may not be accurately known. The marrow dose must be a function of the dose to the blood flowing through the marrow; however, the trabecular formation of bone containing marrow will reduce the dose. The proportion of blood in a given specimen of marrow is poorly known, so that estimates of the dose delivered are inadequately known. Further, any retention of a nuclide in the marrow itself will increase the dose to the marrow correspondingly.

Whenever there is uncertainty as to the sites and lengths of retention for a projected new use of nuclide or method, animal data should first be obtained to indicate the organs most likely to be important during human investigations and to indicate possible unexpected hazards. These results cannot, however, be used to give quantitative values or replace estimates obtainable from experience in man.

D. Estimates of the Dose per Unit of Activity Administered

The dose received by an organ will depend on the activity administered, on the route of administration used and on the fractional uptake of that particular organ.

In general, the nuclide will be administered either orally or intravenously, but other routes may be used, such as intramuscular for clearance studies, and intrathecal for various neurological investigations. Each route of administration requires a knowledge of the physiological pathways and a separate consideration of the dose received.

The data classified in ICRP Publication 2 [2] and in ICRP Publication 10 [3] contribute greatly to this information, but one must remember that these are for healthy adults and that metabolism may be radically changed in particular diseases. Also, these data have not contained information regarding the distribution of radionuclides in pregnant women, foetuses or children. The present revision of the ICRP Committee 2 report will include some data for organic compounds and labelled pharmaceuticals. As has been considered earlier, metabolism may be quite different when labelled pharmaceuticals are used, compared with the metabolism of nuclides in simpler forms.

Another problem has arisen in that the organ may not take up the nuclide or pharmaceutical uniformly. For example, the concentration in the cortex of the kidney may be greater than that averaged throughout the whole kidney. One can conceive that this can be taken to the cellular level and the activity concentrated in groups of cells may require consideration. It is essential therefore that the organ, tissue or structure of reference be clearly defined when making estimates of dose.

Obviously, in these cases, an estimate of the radiation dose likely to be received by a particular patient can only be made on the basis of the best possible information available. When the actual activity has been administered, and measurements of concentration of activity made, then a more accurate assessment of the radiation dose to that particular patient could be undertaken. Considerable discrepancy may arise between the prior estimate and the actual calculated dose, owing to the differences in metabolism and concentration referred to in a previous section.

E. Recommendations on the Clinical Use of Radionuclides

The clinical urgency or importance of the investigation may influence the activity of a nuclide that needs to be used in the investigation of any particular subject. If the activity used is unduly restricted, the diagnostic information obtained from the investigation may, in some cases, be insufficiently exact or detailed for the clinical needs, and the test may then either be wasted or may need to be repeated with a higher activity. This leads to unnecessary irradiation and to a loss of time which may be vital to a patient's health.

1. Categories of Subjects

Bearing these principles in mind, and to help in deciding the activity of a nuclide which should be administered to a particular subject, it is suggested that the following categories should be considered. In this way, general principles and precautions can be stated for each category.

(a) Adult patients who may benefit from the investigation.
(b) Pregnant patients who may benefit from the investigation.
(c) Children who may benefit from the investigation.
(d) Groups of subjects investigated to establish normal values for tests that give abnormal values in certain diseases. These subjects will not themselves necessarily benefit from the test.

2. Basic Principle

The main principle in investigation of subjects is that the activity administered should be the minimum consistent with adequate information for the diagnosis or investigation concerned. This will ensure that the minimum radiation dose is delivered to the patient.

3. Adult Patients

Experience has shown that most tests can be satisfactorily carried out with activities that give rise in adult patients [category (a)] to organ doses of the order of 1 rad, and not usually greater than 5 rads, per investigation. (Tests involving ^{131}I may lead to thyroid doses much greater than this.) However, for a few scanning procedures, it has been found that organ doses of several times these levels are required. Compared with point to point counting, it is necessary with present scanning detecting equipment to use somewhat higher activities to get adequate information on the scan. However, the value of the test to the patient's well-being and/or the seriousness of the disease being investigated may often outweigh any possible long-term radiological hazard, and, in these cases, a higher administered activity may be acceptable.

4. Pregnant Patients

Investigations carried out on pregnant women involve radiation doses to both the mother and the foetus. Consideration must be given to the quantity of activity transmitted across the placenta, and to the resulting foetal uptake. In view of the data from the surveys by MacMahon [4], and by Stewart and Hewitt [5], radiation doses of the order of a few rads may be associated with an increased incidence of leukaemia and childhood malignancies. It is therefore prudent to keep the foetal doses below these levels and to carry out only such investigations that are imperative during pregnancy. The Commission's recommendation (ICRP Publication 9 [6], paragraph 76) regarding the restriction of radiological examinations for women of reproductive capacity should be taken into consideration for the timing of radionuclide investigations.

5. Children

Investigations of children with radionuclides will ordinarily be restricted to those who will benefit from the investigation. The activities to be administered may be calculated approximately by correcting on a weight basis the activity given to an adult, so that the activity administered per kilogram of body weight is comparable. This presumes that the fractional uptake in organs will be similar in children and adults. This will apply for most organs and to most ages of children except for the newborn and children under about one year. In these latter cases, the organ size in relation to the whole body and its uptake must also be considered, as these may be considerably different from those in later childhood. It should be noted that the uptake of bone-seeking nuclides, particularly the alkaline earths, is greater in the growing bone of children than in adult bone. In general, the doses to the organs should be of the same order as (or, if possible, less than) those received by adults during the same investigation. Particular care is required to ensure that the radiation dose received by the gonads is as low as possible, in view of the subsequent child-expectancy of the children.

6. Normal Controls

A requirement for the interpretation of all clinical investigations designed to detect abnormality is that measurements are obtained in subjects who are known to be normal in the relevant respect, in order to establish what are the normal results of the test and their range of variation. This is equally true for investigations involving radionuclides, and hence there is a need for investigations on matched control individuals who may not themselves benefit from the investigation. For investigations in which the standard deviation within normals is small, such volunteer groups can ordinarily be limited to a small number, say 6–10 individuals. The purpose and exact nature of the investigation and the possible hazards should be explained to them before obtaining their consent. The investigations should then be carried out with the minimum activity consistent with obtaining the required information.

The source of such control groups may be individuals attending hospital for other purposes, but if so care must be taken that they are normal in regard to the particular response that is under investigation, which may not be the case if the patients are suffering from another disease. Alternatively, they may be relatives of the patient or members of the general public. There should be reluctance shown in the use of other members of staff or of other groups (such as medical students) if they feel under some obligation to volunteer, and also because they are liable to be used in this capacity by a number of workers independently, without regard to the total dose received by the volunteers.

7. Repeat Investigations

After an investigation has enabled a diagnosis to be made, it is often desirable to test the efficacy of treatment by carrying out a repeat investigation. Similarly, it is sometimes important to carry out serial tests. The overall dose received during each series of tests should be considered, rather than the dose in any one investigation. The use of short-lived nuclides often facilitates serial tests being carried out without the build-up of background activity in the body, which, under some circumstances, necessitates the use of higher activities in the later investigations of the series.

It is inappropriate to specify the number of repeat investigations which should be carried out, because the clinical needs of the patient are of paramount importance. The essential guiding rule is that unnecessary repeat investigations should be avoided.

8. The Use of Blocking Agents

The use of blocking agents (see section B.4), where they are available, should always be considered before administering a radioactive nuclide. An estimate must, however, be made of the dose to the organ which then is likely to receive the maximum dose.

F. Notes on Further Metabolic Information Required for Dosimetry

Besides the information required from the investigation itself, the use of radionuclides and labelled compounds in patients can yield additional scientific knowledge, which can be of great value in assessing the tissue doses involved. Such information, which would be of great importance in the proper use of such tests and to the Commission, may be obtained if investigators, whenever appropriate, secure the maximum information practicable from any investigation, and if this information is subsequently published. The particular data which are not readily available, even though many patients have been investigated over short periods, are the fractional long-term retention of nuclides and labelled compounds, especially those labelled with carbon 14 and tritium and the data on the turnover of particular substances and metabolites. Also, the fraction of orally administered compounds that are absorbed across the gastrointestinal tract is required for normal individuals and for those having particular diseases or surgical resections involving the gastrointestinal tract. Further information on the distribution of radionuclides within organs on the macroscopic and cellular scale is also of great interest. It is only by the collection of more data on the physiological parameters of the metabolism of these compounds and pharmaceuticals that more accurate estimates of the radiation dose to organs, and of its variations in different clinical conditions, may be made.

G. Tabulation of Data

The data tabulated in the last section of this report have been arranged so that each nuclide is dealt with separately. For each nuclide the data are separated according to the different routes of administration and to the different compounds and pharmaceuticals used. For each particular mode of administration the dose received by the principal organs or tissues is given in terms of the dose per unit of activity—usually expressed in millirad per microcurie. The typical activity used in the investigations is usually given where the investigation is in regular use. This statement should not be interpreted as a recommendation of the optimal activity to be administered, but as that which has been found generally necessary in the past. As the sensitivity of detecting systems improves, it should be pos-

sible to reduce these activities, in some cases considerably. For the particular subjects in categories (b), (c) and (d) in section E.1 it will be desirable to reduce the activities administered.

The source of the dose calculations has principally been the open literature, amplified, where necessary, by calculations by the compiler, and by reference to calculations submitted by various workers to the Medical Research Council of the United Kingdom. Their contributions to this work are gratefully acknowledged. Not all the calculations published have been included, for many have been calculated from the same physiological factors; however, a range of values has been included where these exist. It has not been possible to give the physiological data used in each calculation, as these are frequently not included in the publication. It has been assumed that the workers in the field had chosen what they considered were the most appropriate factors.

Where particular physiological data are available, which are not considered to be well known, they have been included in the text.

References to papers containing pertinent dosimetric or physiological data have been given for most nuclides. Again, it has not been the aim to give a bibliography of all references on a particular investigation, and reference may be made to the Nuclear Medicine lists published by the International Atomic Energy Agency and to such sources as Excerpta Medica for such comprehensive bibliographies. During the preparation of this report similar tabulations of data have been compiled in Sweden [7] and in the U.S.A. [8].

References

1. ICRP Publication 16: *Protection of the Patient in X-ray Diagnosis. A report prepared by a task group of ICRP Committee 3.* Pergamon Press, Oxford (1970).
2. ICRP Publication 2: *Recommendations of the International Commission on Radiological Protection —Report of Committee II on Permissible Dose for Internal Radiation (1959).* Pergamon Press, London (1960).
3. ICRP Publication 10: *Recommendations of the International Commission on Radiological Protection—Report of Committee 4 on Evaluation of Radiation Doses to Body Tissues from Internal Contamination due to Occupational Exposure.* Pergamon Press, Oxford (1968).
4. MacMahon, B.: *J. nat. Cancer Inst.* 1962, **28,** 1173.
5. Stewart, A. and Hewitt, D.: *Current Topics in Radiation Research,* Vol. 1. North Holland Publishing Company, Amsterdam (1965).
6. ICRP Publication 9: *Recommendations of the International Commission on Radiological Protection (adopted September 17, 1965).* Pergamon Press, Oxford (1966).
7. National Institute of Radiation Protection, Stockholm: *Stråldoser från radioaktiva ämnen i medicinskt bruk—information till sjukhusens isotopkommittéer* (1969).
8. Hine, G. J. and Johnston, R. E. *J. nucl. Med.* 1970, **11,** 468.

Appendices

Arsenic 74 ^{74}As

Oral

Organ	mrad/μCi	References
GI tract	38	1
Kidney	0.43	1

Intravenous

Typical 1–2 mCi/test.

Organ	mrad/μCi	References
Kidney	14	1
	6	2
Liver	8.4	1

References

1. Vennart, J. and Minski, M.: *Br. J. Radiol.* 1962, **35**, 372.
2. Vennart, J.: PIRC 180(a) of Medical Research Council, U.K. Abstracted from the literature.

Arsenic 76 ^{76}As

Oral

Organ	mrad/μCi	Reference
Kidney	0.09	1

Intravenous ^{76}As

Organ	mrad/μCi	References
Kidney	3 3.2	1 2

References

1. Vennart, J. and Minski, M.: *Br. J. Radiol.* 1962, **35**, 372.
2. Kniseley, R. M., Andrews, G. A. and Harris, C. C.: Editors of *Progress in Medical Isotope Scanning*, Oak Ridge Institute of Nuclear Studies, Tennessee, U.S.A. (TID 7673, 1963), page 375.

Gold 198 ^{198}Au

Oral

Organ	mrad/μCi	Reference
Kidney	0.82	1

Intravenous

Organ	mrad/μCi	Reference
Kidney	8.2	1

Gold 198 Colloid ^{198}Au

Intravenous

Typical 150 μCi/test.

Organ	mrad/μCi	References
Liver	41	2
	44	3
	38	4
	27	5
	40	6
Spleen	37	7
	48	6
Average whole body	1.7	7
Male gonad	0.11	6
Female gonad	0.25	6

Variation of Organ Dose with Age

Organ	Age	mrad/μCi	References	mrad/μCi	References
Liver	Newborn	490	8	380	9
	1 yr	200	8	160	9
	5 yr	120	8	90	9
	10 yr	80	8	70	9
	15 yr	50	8	40	9
	Adult	40	8	30	9
Spleen	Newborn	490	8		
	1 yr	200	8		
	5 yr	120	8		
	10 yr	80	8		
	15 yr	50	8		
	Adult	40	8		

Other Methods of Administration

Treatment Knee Effusions

Typical: Up to 10 mCi. Surface area *circa* 140 cm^2 (refs. 10, 11, 12).
Dose at 1 mm depth 70 rad/μCi deposited over 1 cm^2 ⎫
Dose at 2 mm depth 0.7 rad/μCi deposited over 1 cm^2 ⎬ Deduced from ref. 13.

^{198}Au

Treatment Peritoneal and Pleural Effusions

Typical: Up to 250 mCi.

See Dose Estimates by Walton and Sinclair, 1952 (ref. 13),

e.g. Peritoneal cavity 10 000 cm²

Volume injected (ml)	Average dose to wall from 100 mCi (rad)
100	6 900
300	5 500
500	4 620
1 000	3 260

Endolymphatic Injection

Typical: Up to 60 mCi (ref. 14).

References

1. Vennart, J. and Minski, M.: *Br. J. Radiol.* 1962, **35**, 372.
2. Solomon, A. K., Webster, E. W. and Robinson, C. V.: Report from Harvard Medical School, August 1957.
3. Vennart, J.: PIRC 180(a) of Medical Research Council, U.K. Abstracted from the literature.
4. Kniseley, R. M., Andrews, G. A. and Harris, C. C.: Editors of *Progress in Medical Isotope Scanning*, Oak Ridge Institute of Nuclear Studies, Tennessee, U.S.A. (TID 7673, 1963), page 413.
5. Kniseley, R. M., Andrews, G. A. and Harris, C. C.: Editors of *Progress in Medical Isotope Scanning*, Oak Ridge Institute of Nuclear Studies, Tennessee, U.S.A. (TID 7673, 1963), page 441.
6. McEwan, A. C.: Report NRL/PDS/1967, National Radiation Laboratory, Christchurch, New Zealand.
7. Unpublished calculation submitted to M.R.C. panel.
8. Seltzer, R. A., Kereiakes, J. G. and Saenger, E. L.: *N. Eng. J. Med.* 1964, **271**, 84.
9. Ball, F. and Wolf, R.: *Monatsschrift Für Kinderheilkunde* 1967, **115**, 581.
10. Ansell, B. M., Crook, A., Mallard, J. R. and Bywaters, E. G. L.: *Ann. rheum. Dis.* 1963, **22**, 435.
11. Makin, M. and Robin, G. C.: *J. Am. med. Ass.* 1964, **188**, 725.
12. Virkhunen, M., Krusius, F. F. and Heiskanen, T.: *Acta rheum. scand.* 1967, **13**, 81.
13. Walton, R. J. and Sinclair, W. K.: *Br. med. Bull.* 1952, **8**, 165.
14. Jantet, G. H.: *Br. J. Radiol.* 1962, **35**, 692.

Bismuth 206 ^{206}**Bi**

Subcutaneous

As (a) ^{206}Bi protein, or (b) ^{206}Bi citrate carbon.

Organ	mrad/μCi	References
Kidney: (a) and (b)	56	1
Lymph nodes: (a)	20–300	1
(b)	0.3–4.5	1

Intravenous

Typical 300 μCi/test.

Organ	mrad/μCi	Reference
Kidney	83	2

References

1. Matthews, C. M. E.: *Clin. Sci.* 1962, **22,** 209.
2. Unpublished calculation submitted to M.R.C. panel.
3. Coengracht, J. M. and Dorleyn, M.: *J. belge. Radiol.* 1961, **44,** 485.
4. Mundinger, F.: *Acta Neurochir.* (*Wien*) 1959, Supplement 6, 140.
5. Mundinger, F.: *Nucl.-Med.* (*Stuttg.*) 1959, **I,** 2.

Bromine 82 ^{82}**Br**

Oral

Organ	mrad/μCi	Reference
Average whole body	2.5	1

Intravenous ^{82}Br

Organ	mrad/μCi	References
Average whole body	1.6	2
	2.5	3

References

1. Vennart, J. and Minski, M.: *Br. J. Radiol.* 1962, **35**, 372.
2. Solomon, A. K., Webster, E. W. and Robinson, C. V.: Report from Harvard Medical School, August 1957.
3. Vennart, J.: PIRC 180(a) of Medical Research Council, U.K. Abstracted from the literature.

Carbon 11 Monoxide ^{11}C

Intravenous Injection of ^{11}C as Monoxide in Solution

Typical 30 μCi/test for measurement of red cell volume and blood volume.

Organ	mrad/μCi	References
Average maternal whole body	0.01	1
Average foetal whole body	0.007 7	1
Foetal blood	0.024	1
Placenta	0.055	1

Inhalation

Per breath of 5 mCi held for 20 sec.

Organ	mrad	References
Lungs (apex)	128	2
Blood	211	2
Spleen	109	2
Gonads	61	2

Carbon 11 Dioxide ^{11}C

Inhalation

Per breath of 5 mCi held for 20 sec.

Organ	mrad	References
Lungs (apex)	220	2
Blood	415	2
Spleen	213	2
Gonads	121	2

References

1. Unpublished calculation submitted to M.R.C. panel.
2. Dollery, C. T., Fowler, J. F., Hugh Jones, P., Matthews, C. M. E. and West, J. B.: *Radioaktive Isotope in Klinik und Forschung*, 1963. Verlag von Urban und Schwarzenberg, München.
3. FOWLE, A. S. E., Matthews, C. M. E. and Campbell, E. J. M.: *Clin. Sci.* 1964, **27**, 51.

Carbon 14 Dioxide ^{14}C

Oral and Intravenous

Administration as ^{14}C dioxide in solution.

Organ	mrad/μCi	Reference
Fat	6.3	1

Reference

1. Solomon, A. K., Webster, E. W. and Robinson, C. V.: Report from Harvard Medical School, August 1957.

Carbon 14 Serotonin

14C

Intravenous

Platelet Survival

Typical 4 μCi/test.

Organ	mrad/μCi	Reference
Blood	4	1

References

1. Ellis, R. E.: Unpublished calculations.
2. Leeper, S. W., Brown, H., Davis, V. E., Alfrey, C. P., Kahil, M. E. and Simons, E. L.: *J. clin. Invest.* 1964, **43**, 1296.

Calcium 45

45Ca

Oral

Typical 1–5 μCi/test (usually administered to patients with bone disease or malignant disease).

Organ	mrad/μCi	Reference
Bone	79	1

Intravenous

Typical 1–5 μCi/test.

Organ	mrad/μCi	References
Bone	130	1
	70	2
	58	3
Gonad	0.18	3

Additional Data

Fraction from GI tract to blood in normal = 60% (ref. 4).
Typical values of fraction from GI tract to blood with large calcium loading = 30%.
Typical values of fraction from GI tract to blood in disease-malabsorption = 15%.

References

See ^{47}Ca + ^{47}Sc.

Calcium 47 + Scandium 47 $^{47}Ca + ^{47}Sc$

Oral

Typical 20 μCi/test.

Organ	mrad/μCi	Reference
Bone	19*	1

*$^{47}Ca \sim 95\%$. $^{45}Ca \sim 5\%$.

Intravenous

Typical 5 μCi/test.

Organ	mrad/μCi	References
Bone	31	1
	17	3
	30	5
cf. 19 mrad/μCi for		
$^{47}Ca \sim 95\% + ^{45}Ca \sim 5\%$		6
Gonad	3.1	3
cf. 2.9 mrad/μCi for		
$^{47}Ca \sim 95\% + ^{45}Ca \sim 5\%$		6

Additional Data

Fraction from GI tract to blood in normal = 60% (ref. 4).
Typical value of fraction from GI tract to blood with large calcium load = 30%.
Typical value of fraction from GI tract to blood in disease-malabsorption = 15%.
Calculations must include a correction for 2–5% of ^{45}Ca initially present, the fraction increasing if the solution is stored.

References

1. Vennart, J. and Minski, M.: *Br. J. Radiol.* 1962, **35**, 372.
2. Solomon, A. K., Webster, E. W. and Robinson, C. V.: Report from Harvard Medical School, August 1957.
3. McEwan, A. C.: Report NRL/PDS/1967, National Radiation Laboratory, Christchurch, New Zealand.
4. ICRP Publication 2: *Report of Committee II on Permissible Dose for Internal Radiation* (1959). Pergamon Press, London (1960).

$$^{47}\text{Ca} + {}^{47}\text{Sc}$$

5. Blau, M., Laor, Y. and Bender, M. A.: *Proceedings of IAEA Symposium on Medical Radioisotope Scintigraphy*, Salzburg, August 1968 (1969).
6. McEwan, A. C.: Personal communication.
7. Avioli, L. V., McDonald, J. E., Singer, R. A. and Henneman, P. H.: *J. clin. Invest.* 1965, **44**, 128.
8. Caniggia, A., Gennari, C. and Cesari, L.: *Br. med. J.* 1965, **1**, 427.
9. Kaye, M. and Silverman, M.: *J. Lab. clin. Med.* 1965, **66**, 535.

Chlorine 36 (as perchlorate ion) ^{36}Cl

Oral

Typical 15 μCi/test.

Organ	mrad/μCi	References
Thyroid	24	1
Average whole body	0.2	1

$T_{\text{eff}} = 6$ hours.

Thyroid uptake 10%.

Reference

1. Unpublished calculation submitted to M.R.C. panel.

Chlorine 38 (as perchlorate ion)

Intravenous

Typical 60 μCi/test.

Organ	mrad/μCi	Reference
Average whole body	0.15	1

Reference

1. Unpublished calculation submitted to M.R.C. panel.

Cobalt 56 Vitamin B12

This radiopharmaceutical is not usually used now.

Oral

Organ	mrad/μCi	Reference
Liver	1 400	1

Reference

1. Reizenstein, P.: *Nord. Med.* 1961, **66**, 1618.

Cobalt 57 Vitamin B12 ^{57}Co

Oral

Typical 1 μCi/test.

Organ	mrad/μCi	Reference
Liver	126	1

60% uptake liver. $\qquad$ $T_{\text{eff}} = 156$ days.

Variation of Organ Dose with Age

Organ	Age	mrad/μCi	References
Liver	Newborn	1 500	2
	1 yr	680	2
	5 yr	410	2
	10 yr	280	2
	15 yr	210	2
	Adult	160	2

Schilling Test

After 0.5–1 mg parenteral injection of unlabelled vitamin.
Typical 0.5 μCi/test.

Organ	mrad/μCi	References
Liver	160	3
Male gonad	0.78	3
Female gonad	2.1	3

^{57}Co

Intravenous

Glomerular Filtration Rate

Given after 5 mg blocking dose B12 (liver block).
Typical 2 μCi/test.

Organ	mrad/μCi	References
Kidney	0.3	1
Bladder	1.5	1
Liver	1.3	1
	1.2	4

90 % eliminated in 24 hours (T_{eff} = 4–6 hours).
10 % retained with T_{eff} = 10 days.

Additional Data

Relationship between Quantity of Vitamin B12 administered and Liver Uptake

Quantity of B12 administered	Liver uptake of B12	References
50 μg	3%	5
50 μg	1.7%	6
250 μg	0.7%	6

References

1. Unpublished calculation submitted to M.R.C. panel.
2. Seltzer, R. A., Kereiakes, J. G. and Saenger, E. L.: *N. Eng. J. Med.* 1964, **271,** 84.
3. McEwan, A. C.: Report NRL/PDS/1967, National Radiation Laboratory, Christchurch, New Zealand.
4. Slapak, M. and Hume, D. M.: *Lancet* 1965, **I,** 1095.
5. Glass, G. B. J., Boyd, L. J. and Stephanson, L.: *Science* 1954, **120,** 74.
6. Gaffney, G. W., Watkin, D. M. and Chow, B. F.: *J. Lab. clin. Med.* 1959, **53,** 525.
7. Mollin, D. L.: *Br. med. Bull.* 1959, **15,** 8.
8. Hall, C. A.: *Am. J. clin. Nutr.* 1964, **14,** 156.
9. Nelp, W. B., Wagner, H. N., Jr. and Reba, R. C.: *J. Lab. clin. Med.* 1964, **63,** 480.

Cobalt 58 Vitamin B12 ^{58}Co

Oral

Typical 1 μCi/test

Organ	mrad/μCi	References
Liver	230*	1
	200	2
Average whole body	17	1

*60% absorbed.

Variation of Organ Dose with Age

Organ	Age	mrad/μCi	References
Liver	Newborn	2 300	3
	1 yr	1 200	3
	5 yr	760	3
	10 yr	540	3
	15 yr	420	3
	Adult	330	3

Schilling Test

After 0.5–1 mg parenteral injection of unlabelled vitamin.
Typical 0.5 μCi/test.

Organ	mrad/μCi	References
Liver	210	4
Male gonad	2.6	4
Female gonad	8.3	4

References

1. Unpublished calculation submitted to M.R.C. panel.
2. Reizenstein, P.: *Nord. Med.* 1961, **66**, 1618.
3. Seltzer, R. A., Kereiakes, J. G. and Saenger, E. L.: *N. Eng. J. Med.* 1964, **271**, 84.
4. McEwan, A. C.: Report NRL/PDS/1967, National Radiation Laboratory, Christchurch, New Zealand.
5. Reizenstein, P., Ek, G. and Matthews, C. M. E.: *Phys. Med. Biol.* 1966, **11**, 295.
6. Mollin, D. L.: *Br. med. Bull.* 1959, **15**, 8.

Cobalt 60 Vitamin B12 ^{60}Co

This radiopharmaceutical is not usually used now.

Oral

Organ	mrad/μCi	References
Liver	8 100	1
	6 300	2
	3 000	3
Male gonad	81	1
Female gonad	214	1

Variation of Organ Dose with Age

Organ	Age	mrad/μCi	References
Liver	Newborn	30 000	4
	1 yr	15 000	4
	5 yr	10 000	4
	10 yr	6 900	4
	15 yr	5 300	4
	Adult	4 200	4

References

1. McEwan, A. C.: Report NRL/PDS/1967, National Radiation Laboratory, Christchurch, New Zealand.
2. Vennart, J.: PIRC 180(a) of Medical Research Council, U.K. Abstracted from the literature.
3. Reizenstein, P.: *Nord. Med.* 1961, **66**, 1618.
Also see ^{58}Co and ^{57}Co.
4. Seltzer, R. A., Kereiakes, J. G. and Saenger, E. L.: *N. Eng. J. Med.* 1964, **271**, 84.

Chromium 51 ^{51}Cr

Oral

Stomach Scanning

Typical 200 μCi/test, as sodium chromate.

Organ	mrad/μCi	Reference
Intestine	~5	1

Chromium sesquioxide for faecal marker. Typical 10 μCi/test.
GI tract LLI (normal emptying) 7 mrad/μCi (refs. 2, 3).
Prolonged emptying occurs in some diseased states which leads to doses three times these estimates.

Intradermal Injection

Typical 10 μCi/test.
Activity assumed to be located within a 1 cm sphere (ref. 4).
Average dose to sphere = 2.6 mrad/μCi.

Intravenous

Organ	mrad/μCi	References
As Cr Cl$_3$:		
Kidney	0.56	5
As Na$_2$ Cr O$_4$:		
Blood	2.1	5

Chromium 51 EDTA

Intravenous

Glomerular Filtration Rate

Typical 35 μCi/test (refs. 6, 7).

Intravenous Infusion

Organ	mrad/μCi	Reference
Kidney	0.14	7

Chromium 51 Human Serum Albumen ^{51}Cr

Oral

Typical 50 μCi/test (ref. 8).

Intravenous

Typical 10 μCi/test (ref. 8).

Chromium 51 Labelled Red Blood Cell

Intravenous

Red blood cells labelled *in vitro*   Typical 50–100 μCi/test.
Platelets labelled *in vitro*   Typical 25 μCi/test.
Blood volume labelled *in vitro*   Typical 25 μCi/test.

Red Cell Survival and Blood Volume

Organ	mrad/μCi	References
Spleen	20	9
	70	4
Blood	1.4	10
	2.1	5
	2.4	9
Liver	1.8	11
Gonads	0.22	10
Average whole body	0.25	9

75% Spleen uptake $T_{\text{biol}} = 23\frac{1}{2}$ days (ref. 4).
50% Spleen uptake $T_{\text{biol}} = 33$ days (ref. 4).

Variation of Organ Dose with Age **⁵¹Cr**

Organ	Age	mrad/μCi	Reference	mrad/μCi	Reference
Spleen	Newborn	490	12	600	13
	1 yr	160	12	160	13
	5 yr	100	12	100	13
	10 yr	50	12	60	13
	15 yr	40	12	40	13
	Adult	40	12	40	13
Average whole body	4 mth	1.8	14		
	1 yr 2 mth	1.7	14		
	5 yr	1.2	14		
	6 yr	1.2	14		

Placental Localization

Typical 20 μCi/test.

Organ	mrad/μCi	References
Maternal average whole body dose	0.6	15
	1.0	16
Foetal average whole body dose	0.1	15
	0.2	16

Intrathecal

Typical 50 μCi/test.
The volume is taken to be 150 cc but in some patients it could be smaller (see ¹³¹I references).

Organ	mrad/μCi	Reference
Cells lining ventricles	150	4

Chromium 51 Heat-denatured Red Blood Cells

$51Cr$

Intravenous

Spleen Scanning

Typical 300 μCi/test.

Organ	mrad/μCi	References
Spleen	20	17
	50	10
	10	16
Average whole body	0.36	16
Male gonad	0.1	16
	0.30	10
Female gonad	0.2	16
	0.076	10

Chromium 51 Polystyrene Spheres

Inhalation

^{51}Cr-labelled acetyl acetonate incorporated into 5 μm polystyrene spheres for lung clearance studies. Typical lung burden 1 μCi/test.

Organ	mrad/μCi lung burden	Reference
Lung	10	18

References

1. Griffith, G. H., Owen, G. M., Kirkman, S. and Shields, R.: *Lancet* 1966, **I**, 1244.
2. Whitby, L. G. and Lang, D.: *J. clin. Invest.* 1960, **39**, 854.
3. Pearson, J. D.: *Int. J. appl. Radiat.* 1966, **17**, 13.
4. Unpublished calculation submitted to M.R.C. panel.
5. Solomon, A. K., Webster, E. W. and Robinson, C. V.: Report from Harvard Medical School, August 1957.
6. Sheep, G. F. R., Stacey, B. D. and Thorburn, G. D.: *Science* 1966, **152**, 1076.
7. Garnett, E. S., Parsons, V. and Veall, N.: *Lancet* 1967, **I**, 818.
8. Waldmann, T. A.: *Lancet* 1961, **II**, 121.
9. King, R. and Sharpe, A. R.: *Postgrad. Med.*, September 1963, **34**, A47.
10. McEwan, A. C.: Report NRL/PDS/1967, National Radiation Laboratory, Christchurch, New Zealand.

11. Ellis, R. E.: Unpublished calculations.

12. Seltzer, R. A., Kereiakes, J. G. and Saenger, E. L.: *N. Eng. J. Med.* 1964, **271**, 84.

13. Ball, F., Wolf, R.: *Monatsschrift Für Kinderheilkunde* 1967, **115**, 581.

14. Wellman, H. N., Kereiakes, J. G. and Branson, B. M.: Paper to Symposium on Dose Reduction in Nuclear Medicine, Oak Ridge, Tennessee, December 1969.

15. Paul, J. D.: *Am. J. Obstet. and Gynec.* 1962, **21**, 33.

16. Borner, W.: *Radiologe* 1966, **6**, 323.

17. Kniseley, R. M., Andrews, G. A. and Harris, C. C.: Editors of *Progress in Medical Isotope Scanning*, Oak Ridge Institute of Nuclear Studies, Tennessee, U.S.A. (TID 7673, 1963), page 481.

18. Booker, D. V., Chamberlain, A. C., Rundo, J., Muir, D. C. F. and Thomson, M. L.: *Nature*, 1967, **215**, 30.

19. Mollison, P. L. and Veall, N.: *Br. J. Haemat.* 1955, **I**, 62.

20. Rootwelt, K.: *Scand. J. clin. Lab. Invest.* 1966, **18**, 405.

21. Van Tongeren, J. H. and Majoor, C. L.: *Clin. chim. Acta* 1966, **14**, 31.

22. Jandl, J. H., Greenberg, M. S., Yonemoto, R. H. and Castle, W. B.: *J. clin. Invest.* 1956, **35**, 842.

Caesium 129 ^{129}Cs

Intravenous

Cardiac Scanning

Typical 3 mCi.

Organ	mrad/μCi	References
Liver	0.77*	1
Kidney	2.2†	1
Average whole body	0.23	1

* Uptake 7%.
† Uptake 14.5%.
2 hours for uptake to reach maximum (ref. 2).

References

1. Unpublished calculation submitted to M.R.C. panel.

2. Carr, E. A., Jr., Gleason, G., Shaw, J. and Krontz, B.: *Am. Heart J.* 1964, **68**, 627.

Caesium 130

Intravenous

Cardiac Scanning
Typical 10 mCi.

Organ	mrad/μCi	References
Kidney	0.3	1
Liver	0.04	1

Reference

1. Unpublished calculation submitted to M.R.C. panel.

Caesium 131

Intravenous

Typical 1.25 mCi/test.

Organ	mrad/μCi	References
Liver	0.64	1
Average whole body	0.28	1

References

1. Vennart, J. and Minski, M.: *Br. J. Radiol.* 1962, **35,** 372.
2. Carr, E. A., Walker, B. J. and Bartlett, J., Jr.: *J. clin. Invest.* 1963, **42,** 922.

Caesium 134m

134mCs

Intravenous

Typical 10 mCi/test.

Organ	mrad/μCi	Reference
Average whole body	0.25	1

Reference

1. Unpublished calculation submitted to M.R.C. panel.

Copper 64

64Cu

Oral

Organ	mrad/μCi	Reference
Spleen	0.7	1

Intravenous

Wilson's Disease

Typical 500 μCi/test.

Organ	mrad/μCi	References
Spleen	2.5*	2
Liver	3.5*	2
	1.6	3 (a, b and c)

* 50% at 30 hours in liver, excreta and extracellular space. Remainder may be in kidney.

Copper 64 EDTA

^{64}Cu

Intravenous

Brain Scan for Intracellular Accumulation

Typical 2 mCi/test.

Organ	mrad/μCi	Reference
Spleen	2.5	2

Copper 64 DTPA

Intravenous

Organ	mrad/μCi	References
Average whole body	0.15	4
Liver	1.6	4

References

1. Vennart, J. and Minski, M.: *Br. J. Radiol.* 1962, **35**, 372.
2. Unpublished calculation submitted to M.R.C. panel.
3. Osborn, S. B. and Walshe, J. M.:
   (a) *Clin. Sci.* 1964, **27**, 319;
   (b) *Clin. Sci.* 1965, **29**, 575; and
   (c) *Lancet* 1967, **I**, 346.
4. Kniseley, R. M., Andrews, G. A. and Harris, C. C.: Editors of *Progress in Medical Isotope Scanning*, Oak Ridge Institute of Nuclear Studies, Tennessee, U.S.A. (TID 7673, 1963), page 371.

Copper 67

67Cu

Intravenous

Typical 250 μCi/test.

Organ	mrad/μCi	References
Spleen	6.4	1
Liver	16.8	1

Reference

1. Unpublished calculation submitted to M.R.C. panel.

Fluorine 18 Potassium Fluoborate

18F

Intravenous

Typical 8 mCi $KB^{18}F_4$, with 500 mg perchlorate to stop stomach uptake.

Organ	mrad/μCi	References
Stomach	0.57	1
Female gonad	0.07	1
Male gonad	0.06	1
Bone and teeth	0.6*	1
Bone	0.23	1
	0.2	2
Average whole body	0.07	1
Bladder	2.5–5.0	1

* Also 0.34 mrad/μCi in Na HCO_3 solution.

References

1. Unpublished calculation submitted to M.R.C. panel.
2. Blau, M., Laor, Y. and Bender, M. A.: *Proceedings of IAEA Symposium on Medical Radioisotope Scintigraphy*, Salzburg, August 1968 (1969).
3. Ronai, P., Winchell, H. S. and Anger, H. O.: *J. nucl. Med.* 1968, **9,** 517.
4. French, R. J. and McCready, V. R.: *Br. J. Radiol.* 1967, **40,** 655.
5. Nusynowitz, L., Feldman, H. and Maier, J. G.: *J. nucl. Med.* 1965, **6,** 473.

Iron 55 ^{55}Fe

Oral

Organ	mrad/μCi	Reference
Spleen	2.5	1

Intravenous

Organ	mrad/μCi	References
Spleen	25	1
Blood and heart	4.9	2

The distribution reported is 10% in liver; 15% in bone marrow, with an effective half-life of 2 days. Non-uniform distribution in the liver may give rise to hotspots, however, receiving 100 times the mean liver dose.

References

1. Vennart, J. and Minski, M.: *Br. J. Radiol.* 1962, **35**, 372.
2. Solomon, A. K., Webster, E. W. and Robinson, C. V.: Report from Harvard Medical School, August 1957.
3. Saylor, L. and Finch, C. A.: *Am. J. Physiol.* 1953, **172**, 372.
4. Pirzo-Biroli, G. and Finch, C. A.: *J. Lab. clin. Med.* 1960, **55**, 216.

Iron 59 ^{59}Fe

Ferric citrate, chloride and ferrous ascorbate are used.

Typical 5 μCi/test.

Oral

Organ	mrad/μCi	Reference
Spleen	14	1, 2

contd.

^{59}Fe

Variation of Organ Dose with Age

Organ	Age	mrad/μCi	Reference
Whole body	4 yr	49	3
	5 yr	70	
	6 yr	78	
	15 yr	32	

Intravenous

Organ	mrad/μCi	References
Spleen	140	1, 2
Blood	97	4
Average to tissues (gonad dose estimate)	22	4

Iron 59 EDTA

5 μCi ferrioxamine test.
Large proportion excreted in urine 48 hours (ref. 1).

References

1. Vennart, J. and Minski, M.: *Br. J. Radiol.* 1962, **35,** 372.
2. Vennart, J.: PIRC 180(a) of Medical Research Council, U.K. Abstracted from the literature.
3. Wellman, H. N., Kereiakes, J. G. and Branson, B. M.: Paper to Symposium on Dose Reduction in Nuclear Medicine, Oak Ridge, Tennessee, December 1969.
4. McEwan, A. C.: Report NRL/PDS/1967, National Radiation Laboratory, Christchurch, New Zealand.
5. Fielding, J.: *J. clin. Path.* 1965, **18,** 88.
Also see ^{55}Fe.

Tritium 3 as HTO ^{3}H

Oral

Organ	mrad/μCi	References
Average whole body	0.11	1
Body tissue (43 kg)	0.2	2

$T_{\text{eff}} = 12$ days.

Intravenous

Organ	mrad/μCi	References
Average whole body	0.11	1
Body tissue (43 kg)	0.2	2

References

1. Solomon, A. K., Webster, E. W. and Robinson, C. V.: Report from Harvard Medical School, August 1957.
2. Vennart, J. and Minski, M.: *Br. J. Radiol.* 1962, **35,** 372.

Mercury 197 Chlormerodrin ^{197}Hg

Intravenous

Brain and Kidney Scanning

Typical 0.5–1 mCi/test for brain scans; 50–150 μCi/test for kidney scans.
Great disparity of dose estimates exists due to not taking into account internal conversion x-rays. These earlier estimates are as follows.

Organ	mrad/μCi	References
Kidney	5.4	1
	10	2
	5	3
	2*	4
Average whole body	0.01	1

* With blocking agent.

These are probably too small and should be replaced by the following: **197Hg**

Organ	mrad/μCi	References
Kidney	34*	5
	19	6

* Without blocking agent.

During brain scanning the blocking of the kidney with inactive chlormerodrin results in a reduction of the kidney dose to a half or a third of that without blocking.

Organ	mrad/μCi	References
Male gonad	0.011	6
Female gonad	0.026	6
Ovarian tissue	0.006	7
Ovarian follicle (and developing ovum)	0.140	7

Variation of Organ Dose with Age

Organ	Age	mrad/μCi	References
Kidney	Newborn	170	8
	1 yr	50	8
	5 yr	30	8
	10 yr	20	8
	15 yr	16	8
	Adult	8	8
Kidney	3 yr	68	9
	12 yr	39	9

Additional Data

Concentration in Renal Cortex

Probably twice average kidney dose.
90% kidney uptake $T_{eff} = 0.8$ day.
10% kidney uptake $T_{eff} = 28$ days.

Contamination of ^{197}Hg with ^{203}Hg

Contamination of ^{197}Hg with ^{203}Hg %	Total kidney dose mrad/μCi	References
0	3.5	10
1	3.8	10
2	4.2	10
5	5.2	10
10	6.8	10
100	37.0	10

Mercury 197 BMHP (1 Bromomercury 2 Hydroxypropane)

Intravenous

Spleen Scanning

Typical 150–300 μCi/test.

Organ	mrad/μCi	References
Kidney	20–23	10, 11, 12
	36–70	13

Spleen $T_{eff} = 6$ hours.
Kidney $T_{eff} = 35$ days.

Variation of Organ Dose with Age

Organ	Age	mrad/μCi	References
Kidney	Newborn	400	8
	1 yr	140	8
	5 yr	90	8
	10 yr	50	8
	15 yr	40	8
	Adult	30	8

^{197}Hg

References

1. King, R. and Sharpe, A. R.: *Postgrad. Med.*, September 1963, **34**, A 47.
2. Kniseley, R. M., Andrews, G. A. and Harris, C. C.: Editors of *Progress in Medical Isotope Scanning*, Oak Ridge Institute of Nuclear Studies, Tennessee, U.S.A. (TID 7673, 1963), page 467.
3. Smith, E. M.: *J. nucl. Med.* 1965, **6**, 240.
4. Croll, M. N., Brady, L. W. and Hand, B. M.: *Radiology* 1962, **78**, 635.
5. Pavel, D. and Iionita-Teodorascu, L.: *Oncol. Radiol.* 1967, **6**, 349.
6. McEwan, A. C.: Report NRL/PDS/1967, National Radiation Laboratory, Christchurch, New Zealand.
7. Johnstone, G. S., Cruse, J. R., McIlroy, W. and Kyle, R. W.: *J. nucl. Med.* 1968, **9**, 198.
8. Ball, F. and Wolf, R.: *Monatsschrift Für Kinderheilkunde* 1967, 115 Band, 581.
9. Wellman, H. N., Kereiakes, J. G. and Branson, B. M.: Paper to Symposium on Dose Reduction in Nuclear Medicine, Oak Ridge, Tennessee, December 1969.
10. Radiochemical Centre, Amersham, Buckinghamshire, U.K.: *Technical Bulletins* 67/17 and 67/18 (1967).
11. Kuba, J., Knoll, P., Husak, V., Charamza, O., Wiedermann, M. and Kre, I.: *Z. ges. inn. Med.* 1965, **20**, 431.
12. Wagner, H. N. and Bardfield, P. A.: *J. Am. med. Ass.* 1967, **199**, 202.
13. Borner, W.: *Radiologe* 1966, **6**, 323.
14. Smith, E. M.: *Nucleonics* 1966, **I**, 33.

Mercury 203 Chlormerodrin

^{203}Hg

Intravenous

Brain and Kidney Scanning
^{197}Hg usually used instead.

Organ	mrad/μCi	References
Kidney	86	1
	32	2
	143 (no block)	3
	48 (blocked)	3
	57 (no block)	3
	19 (blocked)	3
	150	4
	76	5
	37	6
Average whole body	0.01	1

Variation of Organ Dose with Age   **²⁰³Hg**

Organ	Age	mrad/μCi	References
Kidney	Newborn	2 480	7
	1 yr	800	7
	5 yr	500	7
	10 yr	310	7
	15 yr	230	7
	Adult	200	7

Mercury 203 BMHP (1 Bromomercury 2 Hydroxypropane)

Intravenous

Spleen Scan

¹⁹⁷Hg usually used instead.
Typical 100 μCi/test.

Organ	mrad/μCi	Reference
Kidney	760	8

References

1. King, R. and Sharpe, A. R.: *Postgrad. Med.*, September 1963, **34**, A47.
2. Kniseley, R. M., Andrews, G. A. and Harris, C. C.: Editors of *Progress in Medical Isotope Scanning*, Oak Ridge Institute of Nuclear Studies, Tennessee, U.S.A. (TID 7673, 1963), page 271.
3. Reference 2, pages 267, 275, 465, and 503.
4. Smith, E. M.: *J. nucl. Med.* 1965, **6**, 240.
5. McEwan, A. C.: Report NRL/PDS/1967, National Radiation Laboratory, Christchurch, New Zealand.
6. Radiochemical Centre, Amersham, Buckinghamshire, U.K.: *Technical Bulletins* 67/17 and 67/18 (1967).
7. Ball, F. and Wolf, R.: *Monatsschrift Für Kinderheilkunde* 1967, **115**, 581.
8. Borner, W.: *Radiologe* 1966, **6**, 323.

Iodine 123

Oral

Organ	mrad/μCi	References
Thyroid	20	1
	16	2
Whole body	0.07	1

Variation of Organ Dose with Age

Organ	Age	mrad/μCi Pure	mrad/μCi +5% ^{124}I	References
Thyroid	Newborn	280	960	3
	1 yr	190	660	3
	5 yr	90	320	3
	10 yr	54	190	3
	15 yr	39	140	3
	Adult	28	110	3

References

1. Evans, K.: M.Sc. Thesis, University of London, 1965.
2. Goolden, A. W. G. *et al.*: *Br. J. Radiol.* 1968, **41**, 20.
3. Wellman, H. N., Kereiakes, J. G. and Branson, B. M.: Paper to Symposium on Dose Reduction in Nuclear Medicine, Oak Ridge, Tennessee, December 1969.

Iodine 124

Oral

Organ	mrad/μCi	References
Thyroid	1 200	1
Average whole body	1 4.2	1

124I

Intravenous

Organ	mrad/μCi	Reference
Thyroid	1 100	2

References

1. Evans, K.: M.Sc. Thesis, University of London, 1965.
2. Vennart, J. and Minski, M.: *Br. J. Radiol.* 1962, **35,** 372.

Iodine 125

125I

Oral

Thyroid Function

Typical 5 μCi/test.

Organ	mrad/μCi	References
Thyroid	1 200	1
Average whole body	1.7–3.8	1

Relative dose with age—see [131]I.

Iodine 125 Labelled Hippuran

Intravenous

Kidney Function

Typical 10 μCi/test.

Organ	mrad/μCi	References
Kidney	0.04	2
Gonad (male)	0.003	2
(female)	0.006	2
Average whole body foetus (from γ rays)	0.002	2

Iodine 125 Labelled Human Serum Albumen 125I

Intravenous

Blood Volume

Typical 5 μCi/test—thyroid gland blocked.

Organ	mrad/μCi	References
Blood	5	3
	1.2	2
Average whole body	2.8	1
Gonad	0.62	2

Placental Localization

Typical 5 μCi ^{125}I HSA/test.

Organ	mrad/μCi	Reference
Average whole body foetus (from γ rays)	0.33/μCi administered to mother	2

Iodine 125 Labelled PVP (Polyvinylpyrrolidone)

Intravenous

Organ	mrad/μCi	References
Liver	41	2
Gonad (male)	0.029	2
(female)	0.099	2

Iodine 125 Labelled Rose Bengal ^{125}I

Intravenous

Organ	mrad/μCi	References
Liver	0.81	2
Gonad (male)	0.001	2
(female)	0.024	2

References
1. Evans, K.: M.Sc. Thesis, University of London, 1965.
2. McEwan, A. C.: Report NRL/PDS/1967, National Radiation Laboratory, Christchurch, New Zealand.
3. Ellis, R. E.: Unpublished calculations.

Iodine 126 ^{126}I

Oral

Organ	mrad/μCi	References
Thyroid	2 300	1
Average whole body	0.9–5.5	1

Reference
1. Evans, K.: M.Sc. Thesis, University of London, 1965.

Iodine 130 ^{130}I

Oral

Organ	mrad/μCi	References
Thyroid	70	1
	186*	2
Average whole body	0.26–0.85	1

* 20% uptake in 20 g gland and biological half-life of 138 days (ref. 3).

References

1. Evans, K.: M.Sc. Thesis, University of London, 1965.
2. Pfannenstiel, P., Sitterson, B. W. and Andrews, G. A.: *J. nucl. Med.* 1968, **9,** 90.
3. ICRP Publication 2: *Report of Committee II on Permissible Dose for Internal Radiation* (1959). Pergamon Press, London (1960).

Iodine 131 (Iodide) ^{131}I

Oral

Thyroid Uptake
Typical 5 μCi.

Organ	mrad/μCi	References
Thyroid	1 200*	1
	1 300	2
	1 800	3
	2 100	4
Parathyroid	330	5
Bladder	2	5
	6.6	6
Kidney	10	5
Stomach	20	5
Salivary gland	50	5
Lactating breast	20	5
Blood	2.5	5
Muscle and bone	<2.5	5
Testis	<2.5	5
Ovary	<1.5	5
	0.3	6
Pituitary	10	5
Liver	10	5
Whole body	0.6–3.6	3
	0.4	6

* Taking into account uptake and thyroid hormone release rates 25 g organ 30% uptake 6.7 d.

     Variation of Organ Dose with Age   **131**I

Organ	Age	mrad/μCi	References	mrad/μCi	References
Thyroid	Newborn	21 000	4	32 000	2
	1 yr	14 400	4	10 000	2
	5 yr	6 900	4	4 300	2
	10 yr	4 000	4	3 100	2
	15 yr	2 900	4	1 700	2
	Adult	2 100	4	1 300	2

Relative Dose with Age (for Iodine 125, 131 and 132)
ref. 7 (deduced from ref. 8)

Age	Thyroid	Whole body	Gonad
Adult	1.0	1.0	1.0
15 yr	1.3	1.2	
10 yr	2.3	1.8	
5 yr	3.3	2.8	
1 yr	7.8	4.5	15
Newborn	23.0	23.0	

Foetal thyroid dose if maternal administration without thyroid blocking.
Foetal thyroid 0–840 mrad/μCi to mother (ref. 9).

Intravenous

Typical 5 μCi for thyroid uptake.

Organ	mrad/μCi	Reference
Thyroid	1 900	10

Iodine 131 Cholografin

Intravenous

Organ	mrad/μCi	Reference
Blood, liver, gall bladder	0.7	11

Iodine 131 Diodrast

Intravenous

Organ	mrad/μCi	Reference
Kidney	1.3	7

Iodine 131 Hippuran

Intravenous

Kidney Scan
Typical 10 μCi.

Organ	mrad/μCi	References
Kidney	0.4	12
Kidney	0.3	1
Kidney (in hypertensive patient)	10	12
Gonad (male)	0.016	1
(female)	0.030	1
Whole body foetus (from γ rays)	0.003	1

contd.   *Variation of Organ Dose with Age*   **131**I

Organ	Age	mrad/μCi	References
Kidney	Newborn	10	2
	1 yr	4	2
	5 yr	3	2
	10 yr	2	2
	15 yr	1	2
	Adult	1	2

Iodine 131 Human Serum Albumen

Intravenous

Blood Volume

Typical 5 μCi/test.   Thyroid gland blocked.

Organ	mrad/μCi	References
Blood	7–28	13
	6.4	1
Whole body	2	13
	1.5	14
Gonads	1.7	1

Variation of Organ Dose with Age

Organ	Age	mrad/μCi	References
Whole body	Newborn	30	15
	1 yr	10	15
	2 yr	12.3	4
	5 yr	7	15
	10 yr	4	15
	15 yr	3	15
	Adult	2	15

Placental Localization

Typical 5 μCi without thyroid blocking.

Organ	mrad/μCi	References
Foetal thyroid	1 000/μCi administered to mother	16
Foetal whole body (average γ-ray dose)	0.7/μCi administered to mother	1

Intrathecal

Typical 100 μCi/test in adult.

If minimum volume 30 ml and maximum $T_{\frac{1}{2}}$ of clearance of HSA = 240 hours, mean dose to superficial layer of spinal cord = 2 000 mrad/μCi.

If volume 100 ml, average $T_{\frac{1}{2}}$ clearance of HSA 4–5 hours, mean dose to superficial layer of spinal cord = 15 mrad/μCi (ref. 17).

Inhalation

Typical 2 mCi/test.

Organ	mrad/μCi	Reference
Lungs	2.1	18

Iodine 131 Heat-denatured Human Serum Albumen

Intravenous

Lung and Liver Scan

Typical 300 μCi/test.

Organ	mrad/μCi	References
Lung	1	19
	6	20
	5	18
Liver	0.3	19
Whole body	0.3	18

Organ	Age	mrad/μCi	References
Liver	Newborn	5	15
	1 yr	2	15
	5 yr	1	15
	10 yr	1	15
	15 yr	<1	15
	Adult	<1	15

Iodine 131 Polyvinylpyrrolidone (PVP)

Intravenous

Brain Scan

Typical 500 μCi/test.

Organ	mrad/μCi	References
Liver	50	21
	36	1
Gonad (male)	0.20	1
(female)	0.38	1

Iodine 131 Rose Bengal

Intravenous

Liver Scan

Typical 200 μCi/test.

Organ	mrad/μCi	References
Liver*	1.1	7
	1–2	22
	2.5	23
	5	1
Gonad (male)	0.032	1
(female)	0.173	1

* *Note:* In patients with cirrhosis or obstructive jaundice the dose to the liver may be greater due to non-excretion. Upper limit 87 mrad/μCi.

contd. *Variation of Organ Dose with Age*

Organ	Age	mrad/μCi	References
Liver	Newborn	10	15
	1 yr	4	15
	4 yr	2.6	4
	5 yr	1.9	4
		2	15
	10 yr	1	15
	13 yr	0.8	4
	15 yr	1	15
	Adult	1	15

Iodine 131 Triolein

Oral

Organ	mrad/μCi	Reference
Body fat	13	7

Variation of Organ Dose with Age

Organ	Age	mrad/μCi	References
Whole body	3 months	7.2	4
	2 yr	2.5	4

Intralymphatic

Triolein in lipiodol fluid.

Organ	mrad/μCi	References
Lung*	10	24
Node	750	24

* Lung uptake = 26% (range of measurements 7%-38%).

References

1. McEwan, A. C.: Report NRL/PDS/1967, National Radiation Laboratory, Christchurch, New Zealand.
2. Seltzer, R. A., Kereiakes, J. G. and Saenger, E. L.: *N. Eng. J. Med.* 1964, **271**, 84.
3. Evans, K.: M.Sc. Thesis, University of London, 1965.
4. Wellman, H. N., Kereiakes, J. G. and Branson, B. M.: Paper to Symposium on Dose Reduction in Nuclear Medicine, Oak Ridge, Tennessee, December 1969.
5. Myant, N. B.: *Minerva nucl.* 1964, **8**, 208.
6. Ellett, W. H.: 1969 Oregon State University (personal communication).
7. Ellis, R. E.: Unpublished calculations.
8. Seltzer, R. A., Kereiakes, J. G., Saenger, E. L. and Myers, D. H.: *Radiology* 1964, **82**, 486.
9. Aboul-Khair, S. A., Buchanan, T. J., Crooks, J. and Turnbull, A. C.: *Clin. Sci.* 1966, **31**, 415.
10. Vennart, J. and Minski, M.: *Br. J. Radiol.* 1962, **35**, 372.
11. Kniseley, R. M., Andrews, G. A. and Harris, C. C.: Editors of *Progress in Medical Isotope Scanning*, Oak Ridge Institute of Nuclear Studies, Tennessee, U.S.A. (TID 7673, 1963), page 451.
12. Laakso, L., Rekonen, A. and Holspainen, T.: *Scand. J. clin. Lab. Invest.* 1965, **17**, 395.
13. Harper, P. V., Siemens, W. D., Lathrop, K. A. and Endlich, H.: *J. nucl. Med.* 1963, **4**, 277.
14. King, R. and Sharpe, A. R.: *Postgrad. Med.*, September 1963, **34**, A47.
15. Ball, F. and Wolf, R.: *Monatsschrift Für Kinderheilkunde*, 1967, **115**, 581.
16. Hibbard, B. M. and Herbert, R. J. T.: *Clin. Sci.* 1960, **19**, 337.
17. Unpublished calculations submitted to M.R.C. panel.
18. Pircher, F. J.: IAEA Symposium Medical Radioisotope Scintigraphy, Salzburg, 1968.
19. Torrance, H. B., Iskister, W. H. and Mitchell, G.: *Br. J. Surg.* 1965, **52**, 813.
20. Furth, E. D., Okinaka, A. J., Fucht, E. F. and Becker, D. V.: *J. nucl. Med.* 1965, **6**, 506.
21. Tothill, P.: *J. nucl. Med.* 1965, **6**, 582.
22. Reference 11, page 274.
23. Reference 11, page 441.
24. Owen, G. M., Crosby, D. L. and Jones-Williams, W.: *Proc. roy. Soc. Med.* 1969, **62**, 545.

Iodine 132   **132**I

Oral

Organ	mrad/μCi	References
Thyroid	20	1
Average whole body	0.08–0.2	1

Variation of Organ Dose with Age   **^{132}I**

Organ	Age	mrad/μCi	References
Thyroid	Newborn	1 200	2
	1 yr	400	2
	5 yr	170	2
	10 yr	120	2
	15 yr	70	2
	Adult	30	2

Relative dose with age—see also ^{131}I.

Foetal Thyroid

Dose if maternal administration without blocking.

Organ	mrad/μCi	Reference
Foetal thyroid	0–70 mrad/μCi administered to mother	3

Intravenous

Placental Localization

Typical 5 μCi ^{132}I HSA without blocking.

Organ	mrad/μCi	Reference
Foetal thyroid	2.6 mrad/μCi administered to mother	4

References

1. Evans, K.: M.Sc. Thesis, University of London, 1965.
2. Seltzer, R. A., Kereiakes, J. G. and Saenger, E. L.: *N. Eng. J. Med.* 1964, **271**, 84.
3. Aboul-Khair, S. A., Buchanan, T. J., Crooks, J. and Turnbull, A. C.: *Clin. Sci.* 1966, **31**, 415.
4. Hibbard, B. M. and Herbert, R. J. T.: *Clin. Sci.* 1960, **19**, 337.

Iodine 133 ^{133}I

Oral

Organ	mrad/μCi	References
Thyroid	430	1
Average whole body	0.7	1

Reference

1. Evans, K.: M.Sc. Thesis, University of London, 1965.

Indium 113m Citrate Complex ^{113m}In

Intravenous

Organ	mrad/μCi	References
Liver	0.55	1
	0.38	2
Spleen	0.16	2

Indium 113m DTPA Complex

Intravenous

Organ	mrad/μCi	References
Blood	0.005	2
Average whole body	0.006	2

See also ref. 3.

Indium 113m Ferric Hydroxide Colloid ^{113m}In

Intravenous

Organ	mrad/μCi	References
Average whole body	0.009 5	4
Blood	0.014	4
Liver	0.45	4
Spleen	0.082	4
Bone marrow	0.018	4

Preparation of colloid (ref. 5).

Indium 113m Ferric Hydroxide Macroparticles

Intravenous

Organ	mrad/μCi	References
Lungs	0.75	6, 7
	0.58	2

References

1. Goodwin, D. A., Stern, H. S., Wagner, H. N., Jr. and Kramer, H. H.: *Nucleonics* 1966, **24**, xi, 65.
2. Unpublished calculation submitted to M.R.C. panel.
3. Stern, H. S., Goodwin, D. A., Scheffel, V., Wagner, H. N., Jr. and Kramer, H. H.: *Nucleonics* 1967, **25**, ii, 62.
4. French, R. J., Johnson, P. F. and Trott, N. G.: *Proceedings of IAEA Symposium on Medical Radioisotope Scintigraphy*, Salzburg, August 1968 (1969).
5. French, R. J.: *Br. J. Radiol.* 1969, **42**, 68.
6. Kramer, H. H. and Stern, H. S.: *J. nucl. Med.* 1966, **7**, 365.
7. Stern, H. S., Goodwin, D. A., Wagner, H. N., Jr. and Kramer, H. H.: *Nucleonics* 1966, **24**, x, 57.

Potassium 43 ^{43}**K**

Intravenous

Organ	mrad/μCi	Reference
Average whole body	1.1	1

Reference

1. Ellis, R. E.: Unpublished calculations.

Krypton 79 ^{79}**Kr**

Inhalation

Cerebral Blood Flow

Organ	mrad/μCi	Reference
Lung	0.1	1

Reference

1. Sokolv, H.: Personal communication.

Krypton 81m ^{81m}Kr

Inhalation

Typical 6 mCi/l breathed for 2 min for lung studies.

Maximum central lung dose 7 mrads/test (ref. 1).

Maximum average lung dose 4.4 mrads/test (ref. 1).

Reference

1. Unpublished calculation submitted to M.R.C. panel.

Krypton 85 ^{85}Kr

Subcutaneous Injection of Saline Solution

Blood Flow in Skin

10 μCi in 0.1 ml distributed subcutaneously in 2.5 cm diameter circle of skin.

Tissue in contact 6.3 rad/μCi (half-life 3 hours) [ref. 1].

Intravenous Saline Solution

(a) *Continuous Drip*

250 μCi/min $\times$ 20 min for cardiac output measurement.

Organ	mrad/μCi	Reference
Trachea	0.8	2

(b) *Single Intravenous Injection*

30 μCi in few ml of saline for cardiac output measurement.

Organ	mrad/μCi	References
Blood	0.1	1
Trachea	0.03	1

Inhalation ^{85}Kr

Cerebral Blood Flow

Typical 0.1 mCi/l air breathed for 14 min.

Organ	1 mCi/l rebreathed for 1 min (mrad)	References
Average whole body	0.6	3
Trachea mucosa	114	3
	114	4
Lungs	29	1

Also used for intra-cardiac shunt with same concentration breathed for 30 sec. $T_{eff} = 12$ min.

Single Inspiration

Typical 10 mCi/test.

Organ	mrad/mCi in one breath	Reference
Trachea mucosa	2.5	1

Determination of Body Fat in Pregnant Women

Typical 20 μCi* rebreathed in a closed circuit spirometer until equilibrium established.

Organ	mrad/μCi	References
Fat	0.014	1 and 5
Water	0.001 4	1 and 5

* 8 μCi remaining in maternal and foetal body at equilibrium which occurs at 90 min in fat person and 60 min in thin person.

References

1. Unpublished calculation submitted to M.R.C. panel.
2. Rochester, D. F., Durand, J., Parker, J., Fritts, H. W. and Harvey, R. M.: *J. clin. Invest.* 1961, **40**, 643.

3. Veall, N. and Vetter, H.: *Radioisotope Techniques in Clinical Research and Diagnosis*. Butterworth and Company, London (1958).
4. Sokolv, H.: Personal communication.
5. Hytten, F. E.: *Proc. Nutr. Soc.* 1964, **XXI**, 23.
6. Lassen, N. A. and Munck, O.: *Acta physiol. scand.* 1955, **33**, 30.

Magnesium 28 ^{28}Mg

Oral

Absorption Studies

Typical 10 μCi/test.

Organ	mrad/μCi	References
37 kg muscle pool	2.3	1
Bone hot spots	1.8*	1

* *N.B.* Various workers have considered that a non-distribution factor of 5 should also be included.   Calculation assumes 85% absorption.

Reports suggest that there may be uptake in cell nucleus (ref. 1).

Intravenous

Typical 20 μCi/test.

Organ	mrad/μCi	References
37 kg muscle pool	2.7	1
Bone hot spots	2.1*	1

* *N.B.* Various workers have considered that a non-distribution factor of 5 should also be included.

References

1. Unpublished calculation submitted to M.R.C. panel.
2. MacIntyre, I.: Magnesium metabolism, *Sci. Basis Med.* (1963).
3. Martin, H. E. and Bauer, F. K.: *Proc. roy. Soc. Med.* 1962, **55**, 912.
4. Silver, L., Robertson, J. S. and Dahl, L. K.: *J. clin. Invest.* 1960, **39**, 420.

Molybdenum 99 ^{99}Mo

Intramuscular

Blood Flow in Muscle
Typical 50 μCi.

Organ	mrad/μCi	Reference
Gonad	0.000 8	1

Intravenous

Typical 50 μCi molybdate.

Organ	mrad/μCi	Reference
Liver	50	2

References

1. Lassen, N. A., Lindbjerg, J. and Munck, O.: *Lancet* 1964, **I,** 686.
2. Sorensen, L. B. and Achanbault, M.: *J. Lab. clin. Med.* 1963, **62,** 330.
3. Rosoff, B. and Spencer, H.: *Nature* 1964, **202,** 410.

Nitrogen 13 ^{13}N

Intravenous in Saline Solution

Organ	mrad/μCi	References
Lung	0.023	1
Gonad	0.000 08–0.000 7	1

Inhalation

Rebreathing N 13 in Closed Circuit

1 mCi/l breathed for 1 min.

Organ	Total dose from 1 mCi/l breathed for 1 min mrad	References
Lung	68	1
Gonad	0.25–1	1
Tissue	0.7	1

Reference

1. Unpublished calculation submitted to M.R.C. panel.

Sodium 22

Oral

Total Exchangeable Sodium

Typical 5 μCi/test.

Organ	mrad/μCi	References
Average whole body	19	1
	32	2
Bone	120*	3

* Assuming 5% retained with T_{eff} = 450 days.

Intravenous

Total Exchangeable Sodium and Retention Studies

Typical 5 μCi/test.

Organ	mrad/μCi	References
Bone	11*	4
Average whole body	17	5

* *Note*: 1% ^{22}Na retained in bone with $T_{eff} \sim$ 1 year assumed in calculations (ref. 4). Otherwise normal T_{eff} = 10 days, but accident cases (burns) T_{eff} up to 20 days (ref. 6).

References

1. Vennart, J. and Minski, M.: *Br. J. Radiol.* 1962, **35,** 372.
2. Veall, N.: *Lancet* 1957, **I,** 653.
3. Veall, N.: *Lancet* 1955, **I,** 419.
4. Miller, H.: *Lancet* 1957, **I,** 734.
5. Veall, N. and Vetter, H.: *Radioisotope Techniques in Clinical Research and Diagnosis.* Butterworth and Co., London 1958.
6. McEwan, A. C.: Personal communication.

^{22}Na

Sodium 24

^{24}Na

Oral

Total Exchangeable Sodium

Typical 20 μCi/test.

Organ	mrad/μCi	Reference
Average whole body	1.7	1

Intravenous

Placental Localization

Typical 8 μCi/test.

Organ	mrad/μCi administered to mother	References
Average whole body	2.0	2
Average foetal whole body	2.8	2
Gonad	1.0	3

References

1. Vennart, J. and Minski, M.: *Br. J. Radiol.* 1962, **35,** 372.
2. Veall, N. and Vetter, H.: *Radioisotope Techniques in Clinical Research and Diagnosis.* Butterworth and Co., London (1958).
3. Lassen, N. A., Lindbjerg, J. and Munck, O.: *Lancet* 1964, **I,** 686.

Niobium 90

^{90}Nb

Intravenous

Organ	mrad/μCi	References
Average whole body	1.7*	1
Bone	3.5*	1
Spleen	4.6*	1
Kidney	5.2*	1
Liver	3.9*	1
Marrow	5.4*	1

* Contaminated with 5% Nb 92 + 1% Nb 95.

Reference

1. Unpublished calculation submitted to M.R.C. panel.

Oxygen 15

^{15}O

Inhalation

Carbon Dioxide

Typical: Up to 4 breaths each of 5 mCi in 1 litre as 1 single breath.

Organ	Total dose from 1 mCi/l single breath (mrad)	References
Lung	16	1
Gonad	3.6	1

Oxygen

Typical 10 mCi in 1 single breath

Organ	Total dose from 1 mCi/l single breath (mrad)	References
Lung	16.6	1
Gonad	3.3	1

Typical test using 8 breaths, each a few millicuries from about 5 mCi/l stock.   **15O**

Organ	Total dose from 1 mCi/l single breath (mrad)	References
Lung	8.4	2
Blood	2.1	2
Gonad	0.56	2

References

1. Unpublished calculation submitted to M.R.C. panel.
2. Dyson, M. A., Hugh Jones, P., Newbery, G. R., Sinclair, J. D. and West, J. B.: *Br. med. J.* 1960, **1**, 230.

Phosphorus 32   **32P**

Oral

Organ	mrad/μCi	Reference
Bone	38	1

Intravenous

Typical 100 μCi/test.

Organ	mrad/μCi	References
Average whole body	10	1
Body tissue	3.5	2
Bone	50	1
	27	2
	20	3
Testes	17*	4

* Mitchell (ref. 4) assumes a concentration in testes of twice that of body concentration.

contd.   *Variation of Organ Dose with Age*   **32p**

Organ	Age	mrad/μCi	References
Bone and marrow	Newborn	550	5
	1 yr	170	5
	5 yr	100	5
	10 yr	60	5
	15 yr	40	5
	Adult	30	5
Liver	Newborn	550	5
	1 yr	170	5
	5 yr	100	5
	10 yr	60	5
	15 yr	40	5
	Adult	30	5
Spleen	Newborn	550	5
	1 yr	170	5
	5 yr	100	5
	10 yr	60	5
	15 yr	40	5
	Adult	30	5

Phosphorus 32 DFP (Diisopropylfluorophosphonate)

Intravenous

DFP-labelled Platelet Survival

Typical 50 μCi/test.

Organ	mrad/μCi	Reference
Blood	50	6

DFP-labelled Red Cell Survival   **32p**

Typical 50 μCi/test.

Organ	mrad/μCi	Reference
Blood	100	7

References

1. Vennart, J. and Minski, M.: *Br. J. Radiol.* 1962, **35**, 372.
2. McEwan, A. C.: Report NRL/PDS/1967, National Radiation Laboratory, Christchurch, New Zealand.
3. Veall, N. and Vetter, H.: *Radioisotope Techniques in Clinical Research and Diagnosis.* Butterworth and Co., London (1958).
4. Mitchell, J. S.: *Br. med. J.* 1951, **2**, 747.
5. Seltzer, R. A., Kereiakes, J. G. and Saenger, E. L.: *N. Eng. J. Med.* 1964, **271**, 84.
6. Leeksma, C. H. W. and Cohen, J. A.: *J. clin. Invest.* 1956, **35**, 964.
7. Ellis, R. E.: Unpublished calculations.
8. Torrance, H. B. and Gwenlock, A. H.: *Clin. Sci.* 1962, **22**, 413 (Colloidal chromium phosphate).

Rubidium 81   **81Rb**

Intravenous

Organ	mrad/μCi	References
Spleen	16.5	1
Average whole body	0.017	1

Including contaminants ^{82}Rb, ^{83}Rb and ^{84}Rb.

Reference

1. Szur, L., Glass, H. I., Lewis, S. M., Grammaticos, P. and Garreta de, A. C.: *Br. J. Radiol.* 1968, **41**, 819.

Rubidium 86 ^{86}Rb

Oral

Organ	mrad/μCi	Reference
Muscle	5.7	1

Intravenous

Organ	mrad/μCi	References
Muscle	5.7	1
	6	2

References

1. Solomon, A. K., Webster, E. W. and Robinson, C. V.: Report from Harvard Medical School, August 1957.
2. Torrance, H. B., Davies, R. P. and Clark, P.: *Lancet* 1961, **II,** 633.

Sulphur 35 ^{35}S

Oral

Organ	mrad/μCi	References
Whole body	2.6	1
Testes	10	1

There is now some doubt as to whether there is preferential uptake of sulphur by the testes.

Intravenous 35S

Typical 50 μCi as sulphate ion.

Organ	mrad/μCi	References
Average whole body	2.6* 0.013†	1 2
Skin	4–6	3
Testes	10*	1

* If 90 day biological half-life assumed.
† If 6 hour biological half-life assumed.

Note: When ^{35}S is incorporated in organic compounds the metabolism of the compound will determine the fate of the ^{35}S label as long as it remains attached, e.g. labelled cystine as an intradermal injection for hair marker and labelled proteins having T_{eff} in the whole body of 14–40 days and globulins $T_{eff} = 63$ days.

References

1. Vennart, J. and Minski, M.: *Br. J. Radiol.* 1962, **35**, 372.
2. Walser, M., Seldin, D. W. and Grollman, A.: *J. clin. Invest.* 1953, **32**, 299.
3. Solomon, A. K., Webster, E. W. and Robinson, C. V.: Report from Harvard Medical School, August 1957.

Antimony 125 125Sb

Intravenous

Organ	mrad/μCi	References
Average whole body	4	1
Liver	50	1
Thyroid	750	1

At end of 24 hours: 30% excreted, 14% thyroid, 56% liver (ref. 1).

Thyroid and liver activities excreted with $T_{biol} = 12$ days.

Reference

1. Unpublished calculation submitted to M.R.C. panel.

Selenium 75 Methionine

Intravenous

Parathyroid Scan

After blocking thyroid with 100 μg triiodothyronine for 7 days.
Typical 250 μCi/test.

Organ	mrad/μCi	References
Average whole body	2.5	1
	3.0	2
	6.0*	3

* $T_{biol} = 100$ days.

Pancreas Scan

Typical 200 μCi/test.

Organ	mrad/μCi	References
Average whole body	8.1	4
	6.5	5
	8.4	6
Pancreas	6.8	5
	0.3	3
	2.5	6
Liver	6.3	5
	0.3	3
	0.8	6
Gonad	10	6
Kidney	56	3

Variation of Organ Dose with Age   **75Se**

Organ	Age	mrad/μCi	References
Liver	7 yr	34	7
Muscle	7 yr	9	7
Pancreas	7 yr	5	7

Foetal Growth Assessment

Typical 1 μCi to mother.

Organ	mrad/μCi	References
Maternal average whole body	10	8
Foetus	18	8

Biological Data

Selenomethionine

Biological half-life 23–36 to 77–144 days.
7% uptake in pancreas 1–2 hours after intravenous administration.
10% uptake in liver.
High concentration at sites of rapid protein syntheses and in placenta and foetuses.

[See ref. 9.]

In adult liver activity constant first day then excreted with a $T_{biol} = 13$ days for the first 10–12 days and then excretion follows a $T_{biol} = 63$ days.

Concentration in muscle reached peak at 10 days.

Whole-body activity followed a $T_{biol} = 130$ days.

Whole-body activity in child followed a $T_{biol} = 70$ days [ref. 8].

Selenium 75 Cystine ^{75}Se

Intravenous

Pancreas Scanning

Typical 4 μCi/kg.

Organ	mrad/μCi	References
Pancreas	2.9	8
Liver	0.4	8
Kidney	53	8
Average whole body	5.3	8

Kidney 50% uptake with $T_{eff} = 10$ hours.

Selenium 75 Selenide

Intravenous

Chondrosarcoma Visualization

Typical 350 μCi/test.

Organ	mrad/μCi	Reference
Liver	13	10

Assumed 10% in liver with $T_{eff} = 43$ days.

References

1. King, R. and Sharpe, A. R.: *Postgrad. Med.*, September 1963, **34**, A47.
2. Kniseley, R. M., Andrews, G. A. and Harris, C. C.: Editors of *Progress in Medical Isotope Scanning*, Oak Ridge Institute of Nuclear Studies, Tennessee, U.S.A. (TID 7673, 1963), page 270.
3. Blau, M.: *Proceedings of IAEA Symposium on Medical Radioisotope Scanning*, Athens, April 1964, Vol. II (1964).
4. McEwan, A. C.: Report NRL/PDS/1967, National Radiation Laboratory, Christchurch, New Zealand.
5. Blau, M.: *J. nucl. Med.* 1961, **2**, 103.
6. Sodee, D. B.: *Nucleonics* 1965, **23**, 78.
7. Wellman, H. N., Kereiakes, J. G. and Branson, B. M.: Paper to Symposium on Dose Reduction in Nuclear Medicine, Oak Ridge, Tennessee, December 1969.

8. Unpublished calculations submitted to M.R.C. panel.
9. Radiochemical Center, Amersham, Buckinghamshire, U.K.: *Technical Bulletin* 67/2, (1967).
10. Modrego, Perez: Personal communication.

Tin 121 $\qquad$ ^{121}Sn

Oral

Incorporation of tin in toothpaste or chewing gum into tooth enamel.
Typical 1 μCi/test.

Organ	mrad/μCi	Reference
Tooth enamel	< 50	1

Reference

1. Unpublished calculation submitted to M.R.C. panel.

Strontium 85 $\qquad$ ^{85}Sr

Oral

Organ	mrad/μCi	Reference
Bone	13	1

Intravenous ^{85}Sr

Typical 26–40 μCi/test.

Organ	mrad/μCi	References
Bone	44	1
	20–50	2
	60	2
	5*	3
	16	4
	44	5
	37	6
	30	7
Average whole body	6	2
	7	4
	23	5
	11	6
Average gonad dose	2.9	3

* This value used metabolic data assuming a 40% long-term retention with a power law excretion of $t^{-0.2}$ and an average geometric factor $\bar{g}$ bone = 20. Reference 1, for example, by comparison uses ICRP Publication 2 [ref. 8] values of 70% retention with a 65 day effective half-life.

Variation of Organ Dose with Age

Organ	Age	mrad/μCi	References
Bone	4 yr	68	9
	10 yr	41	9
	11 yr	63	9
	12 yr	23	9
	18 yr	27	9

References

1. Vennart, J. and Minski, M.: *Br. J. Radiol.* 1962, **35**, 372.
2. Unpublished calculation submitted to M.R.C. panel.
3. McEwan, A. C.: Report NRL/PDS/1967, National Radiation Laboratory, Christchurch, New Zealand.
4. Charkes, N. D., Sklaroff, D. M. and Bierly, J.: *Am. J. Roentgenol.* 1964, **91**, 1121.

5. Spencer, R., Herbert, R., Rish, M. W. and Little, W. A.: *Br. J. Radiol.* 1967, **40,** 641.
6. Meckelnburg, R. L.: *J. nucl. Med.* 1964, **5,** 929.
7. Blau, M., Laor, Y. and Bender, M. A.: *Proceedings of IAEA Symposium on Medical Radioisotope Scintigraphy*, Salzburg, August 1968 (1969).
8. ICRP Publication 2: *Report of Committee II on Permissible Dose for Internal Radiation* (1959). Pergamon Press, London (1960).
9. Wellman, H. N., Kereiakes, J. G. and Branson, B. M.: Paper to Symposium on Dose Reduction in Nuclear Medicine, Oak Ridge, Tennessee, December 1969.

Strontium 87m ^{87m}Sr

Intravenous

Typical 10 μCi/Kg

Organ	mrad/μCi	References
Bone	0.1	1, 2, 3
	0.4	4
	0.6	5
Average whole body	0.02	1
	0.007	2
	0.01	4

Note: In bone calculation

$$\text{Bone } D_\beta = 0.115 \text{ mrad}/\mu\text{Ci.}$$
$$D_\gamma = 0.060\ 6 \text{ mrad}/\mu\text{Ci.}$$
$$D_{\text{total}} = 0.176 \text{ mrad}/\mu\text{Ci.}$$

This ignores factor 5 for non-uniformity, which if included gives 0.636 mrad/μCi.

(These calculations assume 100% in bone; none excreted; and effective half-life = physical half-life and therefore constitutes a maximum dose per μCi.)

References

1. Charkes, N. D., Sklaroff, D. M. and Bierly, J.: *Am. J. Roentg.* 1964, **91,** 1121.
2. Spencer, R., Herbert, R., Rish, M. W. and Little, W. A.: *Br. J. Radiol.* 1967, **40,** 641.
3. Blau, M., Laor, Y. and Bender, M. A.: *Proceedings of IAEA Symposium on Medical Radioisotope Scintigraphy*, Salzburg, August 1968 (1969).
4. Meckelnburg, R. L.: *J. nucl. Med.* 1964, **5,** 929.
5. Radiochemical Centre, Amersham, Buckinghamshire, U.K.: *Technical Bulletin* 68/11 (1968).

Technetium 99m Antimony Sulphide Colloid ^{99m}Tc

Intravenous

Liver Scan

Typical 1 mCi/test.

Organ	mrad/μCi	References
Liver	0.30	1
	0.31	2
	0.33	3
	0.36	4
Spleen	0.14	2
	0.16–0.43	5
	0.22	5
Average whole body	0.017	3
	0.02	1
	0.021	5
Male gonad	0.013	3
Female gonad	0.023	3
Gonads	0.02	5

Variation of Organ Dose with Age

Organ	Age	mrad/μCi	References
Liver	Newborn	2.5	6
	1 yr	1.0	6
	5 yr	0.6	6
	10 yr	0.5	6
	15 yr	0.4	6
	Adult	0.3	6

99mTc

Inhalation

Typical 4 mCi/test.

Organ	mrad/μCi	References
Lungs	0.08	7
Whole body	0.004	7

Additional Data

Preparation of colloid [refs. 8 and 9].

Technetium 99m Human Serum Albumen

Intravenous

Typical 100 μCi.

Organ	mrad/μCi	Reference
Blood	0.05	10

Placentography

Typical 250 μCi or 4 μCi/kg: scanning up to 1 mCi.

Organ	mrad/μCi	References
Maternal average	0.005	11
whole body	0.014	12
Foetus	0.012	12
Maternal blood	0.047	11
	0.045	12
Foetal blood	0.014	11
	0.012	12
Foetal gonad	0.007	13
	0.033	12
Maternal gonad	0.090	12
Foetal thyroid	0.070	12
	0.30*	14
	0.070†	13
	0.050†	10

* If all the 0.4% of the administered activity in the foetal whole body was in the foetal thyroid.
† Unblocked.

Inhalation

HSA Breathed in Bennet Respirator

Typical 1 mCi/test.

Organ	mrad/μCi	References
Average whole body	0.010	11
Stomach	0.060	11
Lower intestine	0.1	11

Additional Data

10–20% retained in lungs; 40% exhaled; remainder swallowed [ref. 11].
HSA labelling method [see ref. 15].

Technetium 99m Heat-denatured Human Serum Albumen

Intravenous

Typical 3–4 mCi/test.

Organ	mrad/μCi	References
Lung	0.22 0.4	4 11

Technetium 99m Iron Complex

Intravenous

Renal Studies

Typical 1 mCi/test.

Organ	mrad/μCi	Reference
Kidney	0.17	16

Retention Data

$T_{\text{eff}} = 6$ hours.

58% in body and of this 10% in each kidney [ref. 17].

Variation of Organ Dose with Age

Organ	Age	mrad/μCi	References
Kidney	Newborn	9	6
	1 yr	3	6
	5 yr	2	6
	10 yr	1	6
	15 yr	1	6
	Adult	1	6

Technetium 99m Pertechnetate

Intravenous

Typical Brain scan   5 mCi in 2–3 ml
       Salivary scan 1 mCi
       Thyroid scan 0.25 mCi (sometimes given orally)

Organ	mrad/μCi	References
Average whole body	0.012	3
	0.013	18
	0.010–0.014	3
	0.012	10
Gonads	0.012	3
	0.014	18
	0.012–0.018	3
	0.016	10
Male gonad	0.012	3
Female gonad	0.018	3
Brain	0.006	19
Salivary gland	0.2–0.5	3
Thyroid	0.1–0.5	3
	0.24	19
Liver	0.032	19
Stomach (intravenous)	0.23	19
(oral)	0.32	19
ULI (brain scan)	0.096	3
(thyroid scan)	0.21–0.29	3
	0.096	10
LI	0.14	19

Cardiac Output

Typical perfusion 2 mCi/ml/min for 5 min [ref. 10].
GI mucosa 2.6 rad total dose.
Blood 0.047 mrad/μCi.

Variation of Organ Dose with Age   **99mTc**

Organ	Age	mrad/μCi	References
Intestine	Newborn	1.6	6
	1 yr	0.4	6
	5 yr	0.3	6
	10 yr	0.2	6
	15 yr	0.1	6
	Adult	0.1	6

Technetium 99m Red Blood Cells

Intravenous

Variation of Organ Dose with Age

Organ	Age	mrad/μCi	References
Spleen	Newborn	20	6
	1 yr	6	6
	5 yr	4	6
	10 yr	2	6
	15 yr	1	6
	Adult	1	6

References

1. Moses, A.: *Picker Scintillar*, October 1966, 18.
2. Herbert, R. J.: *Postgrad. Med.* 1965, **41,** 656.
3. Börner, W.: *Der Radiologe* 1966, **6,** 323.
4. Ellis, R. E.: Unpublished calculations.
5. Smith, E. M.: *J. nucl. Med.* 1965, **6,** 248.
6. Ball, F. and Wolf, R.: *Monatsschrift Für Kinderheilkunde* 1967, **115,** 581.
7. Pircher, F. J.: *Proceedings of an IAEA Symposium on Medical Radioisotope Scintigraphy*, Salzburg 1968 (1969).
8. French, R. J.: *Br. J. Radiol.* 1969, **42,** 68.
9. Lopez, R. B. and French, R. J.: *Br. J. Radiol.* 1969, **42,** 633.
10. Smith, E. M.: *J. nucl. Med.* 1965, **6,** 231.
11. Unpublished calculation submitted to M.R.C. panel.
12. Hibbard, B. M.: Ph.D. thesis, Liverpool University.
13. Herbert, R. J., Hibbard, B. M. and Sheppard, M. A.: *J. nucl. Med.* 1969, **10,** 224.
14. McAfee, J. and Stern, H.: *J. nucl. Med.* 1964, **5,** 936.

99mTc

15. Gwyther, M. and Field, E.: *Int. J. appl. Radiat.* 1966, **17**, 485.
16. Chisholm, G. D. and Aye, M. M.: *Proc. roy. Soc. Med.* 1967, **60**, 869.
17. U.S.A.E.C.: Radioactive Pharmaceuticals Symposium, Oak Ridge, 1965, chapter 18.
18. McEwan, A. C.: Report NRL/PDS/1967, National Radiation Laboratory, Christchurch, New Zealand.
19. Kazem, I., Gelinsky, P. and Schenck, P.: *Br. J. Radiol.* 1967, **40**, 292.

Xenon 133 ^{133}Xe

Intravenous

Typical tests with xenon 133 in saline solution.

Lung investigations   2 injections of 2 mCi each.
Blood flow studies   4 injections of 50 μCi each.
Cardiac output   200μCi in 35 ml saline injected during 5–10 sec.
Blood flow in uterine muscle 1 mCi/test.

Organ	mrad/μCi	References
Lung (normal adult)	0.014	1
Lung (emphysema patient with $\frac{1}{4}$ lung)	0.020	1
Lung (emphysema patient with $\frac{3}{4}$ lung)	0.015	1
Blood	0.04	1
Foetus	0.004	2
Gonad (normal adult)	0.000 1	1
(normal adult)	0.000 4	3
Gonad (emphysema patient)	0.000 3	1

Inhalation

Typical tests using either single breaths of xenon 133 or rebreathing in closed system.

Lung investigations   2 single breaths of 3 mCi with 15 sec holding of breath.
                     2 min rebreathing of concentration of 350 μCi/l to 1 mCi/l repeated twice.
Cerebral blood flow   5 min rebreathing of concentration of 100 to 500 μCi/l.

^{133}Xe

Single Breaths

Organ	mrad/mCi held for 15 sec	References
Lung (normal adult)	43	1
Lung (emphysema patient)	< 43	1
Gonad	0.25	1

Rebreathing

Organ	mrad/1 min rebreathing of concentration of 1 mCi/l	References
Lung (normal adult)	28	1
	35	4
Lung (emphysema patient)	31	1
Fat	9–11	4
Gonad	1.4	1
	2.5–4	4

References

1. Unpublished calculation submitted to M.R.C. panel.
2. Munck, O., Lysgaard, H., Pontonnier, G., Lefevre, H. and Lassen, N. A.: *Lancet* 1964, **I,** 1421.
3. Lassen, N. A., Lindbjerg, I. F. S. and Munck, O.: *Lancet* 1964, **I,** 686.
4. Dollery, C. T., Fowler, J. F., Hugh Jones, P., Matthews, C. M. E. and West, J. B.: *Radioaktive Isotope in Klinik und Forschung.* Verlag von Urban und Schwarzenberg, München (1963).
5. Mallett, B. L. and Veall, N.: *Lancet* 1963, **I,** 1081.
6. Radiochemical Centre, Amersham, Buckinghamshire, U.K.: *Technical Bulletin* 65/12 (1965).

Yttrium 90 ^{90}Y

Oral

Organ	mrad/μCi	Reference
Bone	0.001 9	1

Intravenous

Organ	mrad/μCi	Reference
Bone	19	1

Treatment of Knee Effusions

Typical 100–200 μCi/ml and up to 10–20 ml injected into synovial space.
Dose at 1 mm depth 270 rad/μCi deposited over 1 cm^2 ⎫
Dose at 2 mm depth 80 rad/μCi deposited over 1 cm^2 ⎬ Deduced from ref. 2.

References

1. Vennart, J. and Minski, M.: *Br. J. Radiol.* 1962, **35**, 372.
2. Walton, R. J. and Sinclair, W. K.: *Br. med. Bull.* 1952, **8**, 165.

Zinc 65 ^{65}Zn

Oral

Organ	mrad/μCi	References
Prostate	160	1
Liver	150	1

Intravenous   **^{65}Zn**

Organ	mrad/μCi	References
Prostate	160	1
Liver	150	1

References

1. Vennart, J. and Minski, M.: *Br. J. Radiol.* 1962, **35**, 372.
2. Richmond, C. R., Furchener, J. F., Trafton, G. A. and Langham, W. H.: *Health Physics* 1962, **8**, 481.
3. McKenney, J. R., McClellan, R. O. and Bustad, L. K.: *Health Physics* 1962, **8**, 411.
4. Prasad, A. S., Miale, A., Jr., Sandstead, H. H. and Schubert, A. R.: *J. Lab. clin. Med.* 1963, **61**, 537.
5. Valee, B. C.: *Physiol. Rev.* 1959, **39**, 443.
6. Van Dilla, M. A. and Ergelke, M. J.: *Science* 1960, **131**, 830.